ENVIRONMENTAL CHANGE AND SUSTAINABLE SOCIAL DEVELOPMENT

Environmental Change and Sustainable Social Development

Social Work-Social Development Volume II

Edited by

SVEN HESSLE
Stockholm University, Sweden

Routledge
Taylor & Francis Group
LONDON AND NEW YORK

First published 2014 by Ashgate Publishing

Published 2019 by Routledge
2 Park Square, Milton Park, Abingdon, Oxon OX14 4RN
52 Vanderbilt Avenue, New York, NY 10017

Routledge is an imprint of the Taylor & Francis Group, an informa business

First issued in paperback 2016

British Library Cataloguing in Publication Data
A catalogue record for this book is available from the British Library

The Library of Congress has cataloged the printed edition as follows:
Hessle, Sven, 1941-
Environmental change and sustainable social development: social work-social development/by Sven Hessle.
volumes cm.
Includes bibliographical references and index.
ISBN 978-1-4724-1637-7 (hardback : alk. paper)
1. Social service. 2. Social ecology. I. Title.
HV40.H5747 2014
361—dc23

2013032314

ISBN 978-1-4724-1637-7 (hbk)
ISBN 978-1-138-24713-0 (pbk)

Contents

List of Figures and Tables

Figures

Tables

Notes on Contributors

Francis Adaikalam is Assistant Professor in the Department of Social Work, Loyola College, Chennai, India and is a 'Marie Curie Fellow' under the FP7 International Research Staff Exchange Scheme. He has undertaken numerous action based research on the quality of life of migrants. He received a research fellowship and scholarship from Mahanirban Calcutta Research Group (MCRG), Kolkata and the International Public Health University. He was senior researcher for a research project on a public report on health for the state of Tamil Nadu and team leader for a study supported by UNICEF on the relationship between the mid-day-meal and schooling using the Andaman and Nicobar Island as a case-study. His interests include Social Work Profession, Social Policy, Migration and Public Health.

Fred H. Besthorn holds a M.S.W. and PhD in social work from the University of Kansas and a Master of Divinity from Grace Theological Seminary. He is Professor and MSW Program Director in the School of Social Work at Wichita State University, Wichita, Kansas, USA. He has over fifty published works on a broad array of topics. His work has focused on the development of a conceptual framework for integrating Deep Ecological awareness with social work policy and practice. He has presented at dozens of social work and environmental conferences around the world. His work involves research on the relationship between environmental degradation, sustainability, environmental racism and its social, economic and spiritual impact on disadvantaged and oppressed populations. He is also the creator of *The Global Alliance for a Deep-Ecological Social Work.*

Cíndy Blackstock is Executive Director of the Caring Society and Associate Professor at the University of Alberta in Canada. She is a member of the Gitksan Nation and has worked in the field of child and family services for over 20 years. An author of over 50 publications, her key interests include exploring, and addressing, the causes of disadvantage for Aboriginal children and families by promoting equitable and culturally based interventions. Current professional interests include holding fellowships with the Ashoka Foundation and the J.W. McConnell Family Foundation.

Alejandro Bonilla Garcia is from Mexico where he completed his Actuarial studies at the Anahuac University. He holds a MSc Degree in Applied Mathematics to the Economy as well as a PhD in Economics from the University of Social Studies of Toulouse, France. He is Senior Actuarial Adviser to the Deputy Director General for Policy of the ILO in Geneva Switzerland. He has been Director a.i. of the Social Security Department of the ILO; Secretary General a.i. of the International Social Security Association; Deputy Regional Director of the ILO for the Americas; Social Security Expert for Central America and the Caribbean and Senior Actuary of the International Actuarial Service of the ILO. He founded and was an associate of two consultancy firms. Economic and Actuarial Consultants, (CEASA) S.A. and Project Development of the Northwest (DIPNOR). Mr. Bonilla Garcia was also the developer and first Director of the Sonora Investment Fund (FISON). Mr. Bonilla Garcia has numerous publications and has participated in fora all over the world. He is member of the Mexican College of Actuaries (CONAC), member of the Mexican Association of Actuaries (AMA) and member of the Swiss Association of Actuaries.

Monit Cheung MA, MSW, PhD, LCSW is Professor and Chair of Clinical Practice Concentration at the Graduate College of Social Work, University of Houston, USA. She is Principal Investigator of the Child Welfare Education Project for training child welfare social workers and Associate Director of the Child and Family Center for Innovative Research. She is a Licensed Clinical Social Worker specializing in play therapy, family counseling, child/adolescent counseling, child protection, sexual and domestic violence, and incest survivor treatment. A volunteer clinician by counseling and case consultation at the Asian American Family Services, and a consultant trainer for the Hong Kong Social Welfare Department and the Hong Kong Police Force. She has presented in numerous workshops and conferences and written numerous publications on child protection and parenting issues in English and Chinese.

Elizabeth J. Clark, PhD, ACSW, MPH, is Chief Executive Officer of the National Association of Social Workers-USA. Located in Washington, DC, the NASW is a membership organization of professional social workers with 140,000 members. In addition to promoting, developing, and protecting the practice of social work and social workers, NASW has a strong social justice and advocacy mission. She is a member of the Academy of Certified Social Workers and the National Academies of Practice. Dr. Clark's background is health care with a specialization in oncology. She has worked in direct clinical practice; has held several academic positions including an Associate Professorship in Medical Oncology; has been a hospital administrator; has conducted research and has published in areas such as cancer survivorship, hope, and bereavement.

Antoinette Y. Farmer PhD, University of Pittsburgh, USA. Associate professor and associate dean for academic affairs in the school of social work at Rutgers, The State University of New Jersey, USA. Her research focuses on examining the social and interpersonal factors that affect parenting as well as how parenting practices influence adolescent high risk behaviours, such as delinquency and substance use. She has published widely and co-edited a special issue of the

Journal of Social Service Research, on conducting research with minority and oppressed populations.

Katie Findley MSW, is a PhD student in the school of social work at Rutgers, The State University of New Jersey, USA. Her research interests include human trafficking for organ removal, domestic and international organ transplantation policies, chronic kidney disease, and international health inequalities. She is a graduate research assistant at the Rutgers University Institute for Families where she engages in child support research and training evaluation. Katie is engaged in research, evaluation, and collaboration with international non-governmental organizations and educational institutions in China to improve and increase social services and social entrepreneurship as a research assistant for the Huamin Research Center. She also is a Delegate Assembly member for the state of New Jersey for the National Association of Social Workers.

Sven Hessle PhD, is a Psychologist and Senior Professor of Social work at Stockholm University. His research concerns the study of poverty, families and children in an international perspective as well as Comparative International social work in general. He has published alone, co-edited or contributed in numerous books, research reports and articles. He is founder and Editor-in-Chief of the *International Journal of Social Welfare*. He was given the Katherine Kendall Award from IASSW in August 2006 for his distinguished contributions to international social work education.

Elizabeth Hoffler MSW, ACSW, is special assistant to the chief executive officer and senior policy analyst at the National Association of Social Workers-USA. She advises the CEO on key planning, strategic, and policy issues, and assists in managing special projects within the association. Additionally, she works within the advocacy and social policy department, supervising Political Action for Candidate Election (PACE) which endorses federal candidates for election, and the manager for the Social Work Reinvestment Initiative. Ms. Hoffler also oversees NASW's efforts on behalf of service members, veterans, and their families, most recently managing an initiative in collaboration with the White House Joining Forces Initiative.

Laurel Laiewski MSSW, is a doctoral student at the University of Tennessee (UTK) College of Social Work, Knoxville, Tennessee, USA. She received her master's degree in social work from UTK with a management and community practice concentration and her bachelor's degree in psychology from Clemson University in Clemson, South Carolina. She has worked and volunteered for several non-profits in the Knoxville, Tennessee for the past five years. She studies displaced populations, refugee resettlement in the United States, the socio-cultural implications of climate change, disaster policy, and social work and the natural environment.

Antoinette Lombard is professor in Social Work and head of the Department of Social Work and Criminology at the University of Pretoria in South Africa. She is actively involved in social work education and professional development of social work in South Africa. She is the international representative of the Association of South African Social Work Education Institutions (ASASWEI) and in this capacity a board member of the International Association for Schools of Social Work (IASSW). Her research interests include developmental social work with vulnerable groups, social and economic development and policy and more recently, environmental and green social work.

Thomas Matyók PhD is Associate Professor, Chair in Department of Peace and Conflict Studies, University of North Carolina at Greensboro, USA. His research interests include the education and training conflict workers, the changing nature of war, violence, modern day slavery, and human and environmental oppression. He has extensive experience in crisis negotiation and crisis management. He is presently engaged in research on international curriculum development for global citizenship. His educational and scholarship expertise involve creating responses to complex, global issues including war, poverty, and violence, which result in the oppression and degradation of both the community and the environment.

Pius Mutuku Mutie holds a BA degree in social work and an MA degree in sociology from the University of Nairobi and a Ph.D. in social anthropology from the University of Bielefeld in Germany. He is currently Senior Lecturer in the Department of Sociology and Social Work, University of Nairobi. He is also the coordinator of the social work studies programme. Dr Mutie has taught social work at the University of Nairobi since 1993. He has published in the disciplines of social work, sociology and anthropology. Dr Mutie is also a consultant on social/community development issues.

Rhonda G. Patrick LCSW, MPA, is a social work educator, research and practitioner. She is a Licensed Clinical Social Worker, with over 20 years of clinical, administrative, and governmental social work practice experience. She provides executive consultation services to community based organizations primarily regarding program design, implementation and evaluation, revenue development, and business administration. She is also the CEO of a behavioural health organization that manages large governmental contracts for smaller organizations. In addition to working in business and administrative enterprises, she is the primary researcher and content designer who owns and operates a web-based learning management company with over 4,500 active students. Ms. Patrick focuses her research on the use of technology in direct social work practice.

Ronald Quincy PhD is director of the Center for Nonprofit Management and Governance, School of Social Work, Rutgers, The State University of New Jersey, USA where he teaches non-profit and public management courses. He is co-editor-in-chief of the Journal for Nonprofit Management. Ron's research interest includes disaster relief, and institutional leadership and governance. Ron served as a cabinet member under two State of Michigan governors; was White House Fellow/Special Assistant to the Secretary, Housing and Urban Development; Foreign Policy Advisor, US. State Department; Associate Vice President and Assistant to the President, Harvard University; Executive Director, Martin Luther King Center for Nonviolent Social Change; Executive Director of the Congressional Black Caucus Foundation; and President, White House Fellows Foundation.

Mary E. Rogge MSW, PhD is associate professor at the University of Tennessee (UTK) College of Social Work, Knoxville, Tennessee, USA. She studies environmental justice for children, displaced persons and other vulnerable populations, disaster management and community-based participatory research. She is co-principle investigator of the interdisciplinary UTK Youth, Environment, and Health research team. On the board of the International Consortium for Social Development. Chairperson and treasurer of the managing board of *Social Development Issues*, Chairperson of the U.S. Association for Community Organization and Social Administration. Currently a Commissioner on the Global Social Work Education Commission of the U.S. Council of Social Work Education and a board member and fellow of the UTK Center for the Study of Social Justice.

Sarangdhar Samal heads Orissa's leading social developmental organization (NYSASDRI) in India. He has been involved in action and research development for twenty five years. He has pioneered and initiated a number of change models in rural Orissa. He is the Founder of Kalinga Eye hospital which caters to a population of four million of the most deprived areas of Orissa on both a low and no cost basis. He recently founded the Nysasdri School of Journalism and Social work to offer two years master's degrees in social work and journalism.

Cathryne L. Schmitz MSW, PhD is Professor in the Department of Social Work and and Chair in Department of Peace and Conflict Studies, University of North Carolina Greensboro, USA. She has extensive experience in the fields of leadership, community building, and macro practice. Much of her scholarship focuses on organizational and community change, critical multiculturalism, privilege/oppression, leadership, multi-disciplinary education and practice, global engagement, and environmental concerns, responses, and education. She has numerous publications, including a co-authored book on *Critical Multicultural Social Work*. She is actively engaged in global and environmental education and scholarship having presented nationally and internationally on these issues.

Roland J Sewell MBE BSc (Hons) MICE CEng, entered ministry with The Salvation Army, and has worked for decades in relief and development programs. Lt. Colonel Sewell and his wife founded The Salvation Army's International Emergency Services. He has responded to complex emergencies in times of war and peace with programmes as diverse as cholera preparedness, feeding, psycho-social support, water supply and rebuilding homes, lives and livelihoods. He admires the strength of the human spirit to recover after loss. In retirement he has founded *LetsMakeItHappen Ltd*; a coaching enterprise, engaging with, motivating and inspiring individuals and communities to overcome challenges and develop good solutions. In 1992 he was honored by Queen Elizabeth II with the Order of the British Empire for services to The Salvation Army and OXFAM.

Decha Sungkawan is Associate Professor of Social Work and Criminal Justice, and Dean at the Faculty of Social Administration, Thammasat University, Thailand where he teaches social work, social policy and social welfare, criminology and criminal justice. His research interests and publications have addressed the interrelated issues of social work and social welfare and criminal justice. He has been involved in several studies and development projects in and outside Thailand. He has his first degree in Political Science from Thammasat University and holds PhD from the School of Social Service Administration, the University of Chicago, USA.

Deirdre Tedmanson is Lecturer in the School of Psychology, Social Work and Social Policy at the University of South Australia, and Associate Researcher with the Hawke Research Institute and a Research Scholar with the Centre for Aboriginal Economic Policy Research at the Australian National University. Deirdre has published widely on a range of subjects including Indigenous community and enterprise development, women's empowerment, and participatory action research. Her research and teaching interests include the critical analysis, organizational, management, and policy dimensions of human service delivery, particularly for Indigenous Australians. A major focus of Deirdre's research is working in collaborative partnerships with people and organizations from the A̲nangu Pitjantjatjara Yankunytjatjara (APY) Lands of central Australia, on A̲nangu initiated social and emotional well-being, community development and social enterprise initiatives.

Hilary N. Weaver (Lakota) is a Professor and Associate Dean for Academic Affairs in the School of Social Work, at SUNY-Buffalo, USA. Her teaching, research, and service focus on indigenous populations. She currently serves as President of the American Indian Alaska Native Social Work Educators Association. Dr. Weaver has presented her work regionally, nationally, and internationally including presenting at the Permanent Forum on Indigenous Issues at the United Nations. She has numerous publications, and is currently compiling an edited book, *Social Issues in Contemporary Native America: Reflections from Turtle Island.* Dr. Weaver has received funding from the National Cancer Institute to develop and test a culturally-grounded wellness curriculum for urban Native American youth, the *Healthy Living in Two Worlds* program.

Tom Vickers is Lecturer in Social Sciences at Northumbria University (UK). During 2011 and 2012 he was employed at Durham University on the ESRC-funded project 'Internationalising Institutional and Professional Practices: Community Partnership Models of Change in Post-Tsunami Sri Lanka'. A related set of themes run through Tom's research, concerning: the impact of international, national and local economic and political contexts on policies affecting migrants and ethnic minorities; the interaction between policy, professional practice, and community action; His current work focuses on the impact of the economic crisis and the British government's 'austerity programme' on refugees, asylum seekers and Eastern European migrant workers in North East England.

Andries Viviers is a graduate student of the University of Pretoria and Senior Social Policy Specialist for UNICEF South Africa. He works actively in the field of child rights and is particularly interested in promoting the participation of children in social dialogues through ethical participation methods and ensuring that information in the public domain is accessible for children. He has contributed to many child rights-related publications and was member of the editorial team of the *South African Child Gauge* 2010/2011 that focuses on children's participation.

Xiying Wang is Associate Professor in the school of Social Development and Public Policy at Beijing Normal University, China. She teaches qualitative research methods, feminist theories, and disaster social work. Her research interests include resilience, HIV/AIDS, violence against women, and disaster social work.

Xiulan Zhang obtained her PhD degree in social welfare from the University of California at Berkeley, USA. She is Professor and the Dean of the School of Social Development and Public Policy, Beijing Normal University, China. She is on the national advisory councils of healthcare reform, poverty alleviation, and aging; is a member of the Advisory Committee of the Ministry of Civil Affairs; is the vice-president of China Association of Social Welfare and China Social Policy Association in China. She has published extensively in the areas of education, social protection, social welfare, community participation, long term care, poverty and illness, early child care and education, health care and housing issue of the migrant populations, child protection and child poverty, etc. She has consulted to the Ministry of Civil Affairs, the World Bank, the Asian Development Bank, UNICEF, DFID, the Ford Foundation, and other agencies.

Chapter 1
Introduction

Sven Hessle

A Global Agenda as Point of Departure – Development of Themes

This is the second volume of three covering the most important aspects of social work and social development themes in the world today. *Environmental Change and Sustainable Social Development* is the topic in focus in this volume. The other Volumes cover *Human Rights and Social Equality: Challenges for Social Work (Volume I)* and *Global Social Transformation and Social Action: the Role of Social Workers (Volume III)*.

The main international organisations representing social policy and social work gathered together in Stockholm in July 2012 under the main theme *Social Work and Social Development 2012: Action and Impact.*[1]

At the 2012 conference, we wanted to advance the work on the Global Agenda and beyond by addressing and demonstrating the actions that might be required to develop and accomplish the Agenda – with regard to methods in practice and research, in social policy and social work education, and in a broader discourse of global commitment and cooperation.[2]

The conference also provided opportunities to discuss issues which the social work and social welfare sectors face every day and to ensure a closer link between evidence-based practice, policy objectives and social development goals. Furthermore, the conference revealed how the actions that were meant to develop and accomplish the Global Agenda, might impact on the conditions of people and demonstrate how actions in the social work and social development community can contribute to an enduring physical environment and sustainable social development.

Selection of Contributions for the Volumes

It is an understatement to say that the Stockholm conference was a success with about 2,500 participants representing 106 nations present. Around 2,100 abstracts

1 The following three basic organisations were responsible for the conference. International Associations of Schools of Social Work (IASSW); International Council on Social Welfare (ICSW) and International Federation of Social Workers (IFSW).

2 The Global Agenda with the commitments for the three international organisations is published as appendix in this volume.

and poster suggestions were received in advance for assessment by an international selected panel. Every contribution was evaluated on a Five-point scale by two independent members of the panel. It should be noticed that we did our best to recruiting expertise in the panel for the three main themes of the conference.

The volumes that comprise the three themes have been edited with the intention of being representative of the basic theme covered, including a wide panorama of international examples of implementations. But, first and foremost, *the contributions were selected from those abstracts that the international panel evaluated as the best within its theme.*

We could not cover all the subjects in the conference completely, but the priority was to deliver academic research presentations as well as presentations from practicing social workers.

I am sure that any reader, whether from social policy/welfare, social work practice, education or research will find treasures for inspiration in this kaleidoscope of current implementations and discourses round the world.

Environmental Change and Sustainable Social Development – an Overview

With global warming and the increase in natural disasters, an alternative approach to the natural environment is vital. The main focus of this volume is to emphasise the person-in-environment concept and to find measures for implementation of it. For social work the environment has traditionally been viewed as a world of human relationships. The physical world has had a hard struggle to enter the environmental perspective. The contributions from indigenous people wherever they live, can teach us how to live with, have a dialogue with and appreciate nature. The increasing number of disasters, manmade or natural, seems to be nature's answer to international neglect.

In this volume we have the pleasure of presenting theoretical and practical platforms for *Environmental Social Work* or *Green social work*, or whatever you want to call what we call a *new paradigmatic shift* in our attitude to *person-in-environment.*

The representatives of this paradigmatic change in social work have written for this volumeand together with the contributions in the special issue on environmental social work in the *International Journal of Social Welfare* that was published for and distributed to all participants at the Stockholm Conference, would be enough knowhow to make a difference in the future.[3]

In this volume we have collected lots of ideas about how social work has developed the profession in different forms of Disaster Management related to manmade or natural disasters around the world.

A final question is what actions are needed to integrate the three pillars of Economic development, Environmental development and Social protection?

3 Int J Soc Welfare 2012:22: July.

Some contributions might be of inspiration for challenging this question, posed in 1992 at the UN Summit in Rio de Janeiro, especially in Chapters 6, 8 and 9.

Cindy Blackstock (Canada) opens this volume in Chapter 2 with an important message: *The Government of Canada is on trial for racial discrimination against First Nations children.* The systematic deprivation of proper education coupled with the trauma related to cultural dislocation, abuse and neglect spurred multi-generational cycles of trauma and poverty that continue to undermine First Nations families today. While First Nations are confident that the Tribunal, upon hearing all of the evidence during 2013, will conclude that discrimination has occurred, there is no doubt that this case will set an important national and international precedent on Indigenous child rights.

In Chapter 3, *Environmental Social Work*, Fred Besthorn (USA) commands those who are representing social work to join with those peoples at local levels most impacted by environmental decline and most knowledgeable and prepared to take the action necessary to improve their unique situations.

In his own words:

> It must begin with our genuine curiosity for what is possible, must be imbued with a spirit of contemplation, considered thought and deliberation concerning the next steps and must be sustained by a commitment to connection, bilateral and multilateral cooperation, and inclusivity of the broadest possible range of voices and perspectives. And finally it must fully incorporate the experiences, insights and real-world solutions that every day, hard-working people have into what is important and necessary to their well-being – in their place and at their time. This is what social work does, it is what we can do well and it is what we must do now.

In Chapter 4, *Disaster Management: Perspectives for Social Work and Social Development*, we look into what disaster management means, both theoretically and practically, for different disasters in different parts of the world.

Roland Sewell (UK) brings us closer to the concept through the *Disaster Management Cycle* for planning for; reducing the impact of reaction during and immediately following and then recovery from a disaster. Then follows a few examples of disaster management in different part of the world, and where social work has had to fight for, but found, its acknowledged role.

One key to that accomplishment is formulated by *Xiulan Zhang* (China) through the experience of Chinese social work action and impact in disaster management of the earthquakes that hit China on a regular basis:

> the acts of persistence and companionship that qualify social workers as helping professionals and, arguably more important, wins the hearts of the people.

Decha Sungkawan (Thailand) includes a Gender Awareness Perspective in the guidelines for social workers from both public and private sectors, from the established *Manual for Disaster Management* in Thailand which has been stricken by many flooding disasters during recent years.

The 2004 South Asian tsunami affected 12 countries. In Sri Lanka, immediate aid was provided primarily by local people. We present two of the still ongoing studies from the aftermath of the disaster. Tom Vickers (UK) and Lena Dominelli (UK) specify two different models for empowering interventions involving intensive and long-lasting partnerships built on trust and listening to local actors in order to engage local communities in the empowering processes of giving and receiving aid.

Lena Dominelli has compiled qualitative data with a special focus on women's experiences. She found that stereotyped, unitary identities and traditional gender relations excluded women, during and after the tsunami. Realising equality and empowerment for women in such circumstances is difficult, and makes listening to women crucial to aid responses.

The more recent earthquake that struck Haiti, killing 316,000 individuals, injuring 300,000, and causing one million people to lose their homes, caused The State University of New Jersey in the US to be first social work responders in the aftermath of the earthquake. Antoinette Farmer (USA) and Ronald Quincy (USA) describe the training, including the field experience, social workers are provided with.

Chapter 5, *Indigenous Understanding of Environmental Change and its Social Consequences*, presents two contributions from different parts of the world.

In the first part, Hilary Weaver (USA), a First Nation Lakota, gives the reader a thorough understanding of how climate change is influencing the lives of indigenous people. Her own Lakota teachings tell her that: ' ... Seven generations ago my ancestors planned for me in ways that assured a place for me in this contemporary world. I have the same responsibility to plan for the next seven generations'.

In the second part Sarangadhar Samal (India) communicates the success story about how, over the years, a fishing village in India that was devastated during a flood, became a green, living and prosperous, local agricultural community.

Chapter 6, *Integrating Economic, Environmental and Social Perspectives*, discusses the current challenges after all the UN summits conducted since 1992. Four contributions are presented: The ILO's Social Protection Floors Recommendation; the interesting case of rapid Rwandan development ; children's voices as agents for the right to survive climate change and finally, an example of the unavoidable multidisciplinary approach for getting close to a holistic vision for action.

Alejandro Bonilla Garcia gives a detailed presentation of the newly adopted Recommendation on The Social Protection Floor, R 202, by the International Labour Organization (ILO). It is clear in stating that social protection floors are a fundamental element of national social security systems and that they should be established and maintained within strategies for the extension of social security that progressively ensure higher levels of social security to as many people as possible, guided by ILO social security standards.

Antoinette Lombard (South Africa) and André Viviers (South Africa) describe from their South African /UNICEF collaborative project how important it is to include children as active citizens in social, economic and environmental development. Globally, climate change will have a profound impact on children's right to survival, development, protection and participation.

In Multidisciplinary education for environmental sustainability, Cathryne Schmitz (USA) and Tom Matyók (USA) describe the development of a fascinating multidisciplinary course in which students work at the micro-, meso-, and macro-levels through the lens of mega analysis. At the core of the educational process is multidisciplinary team-building in which students have a collective responsibility for their learning. Through this process students are immersed in a culture that values civic engagement and global citizenship.

Chapter 7, *Social Action for Clean Water* is of course one of the future basic issues has two contributions calling for action.

Pius Mutuku Mutie (Kenya) notes that only 57 per cent of households in Kenya have access to water that is 'considered safe'. And the proportion is even lower (only 50%) for rural populations. There are many households in Kenya whose nearest water source (a river, stream, well or a dam) is more than 10 kilometres away. So the priority in many arid and semi-arid parts of Kenya has actually been one of access to 'just' water and not necessarily 'clean' water. Through a bottom-up and participatory approach, social workers are pivotal in preparing communities to engage in sustainable projects. In a poor country such as Kenya, community organisation/development skills are necessary to enable a social worker to know how to mobilise communities for larger development programmes and empower the people to self-determination.

Exposure to chemicals has been documented to have harmful physical effects on children and adults. With a focus on children, Katie Findley (USA), Laurel Laiewski (USA) and Mary E. Rogge (USA) show, with much empirical evidence, that children are especially at risk from and sensitive to pesticide exposure. The authors provide information with resources about policy, practice and research on pesticides and water at local, national and international levels.

In Chapter 8, *Social Economy and Sustainable Social Development: Local Lessons: Global Challenges*, Deirdre Tedmanson (Australia) presents an enterprise development project in collaboration with remote indigenous communities in Australia. Community members operate an academic, cultural tourism venture entitled '*Learning Country*' which aims to enable social work academics and students to participate in an 'on country' pre-placement learning exchange experience, hosted and guided by elders and leaders from the Anangu homelands.

Francis Adaikalam (India) takes us to the world of microfinance systems in a comparative study of rural housing districts in India.

In Chapter 9 *Sustainable Social Development: Research, Education and Practice*, Rhonda Patrick (USA) and Monit Cheung (USA) present a design for a global access education model using technology to connect both urban and rural universities from six continents. Twelve universities have been invited to participate in this project. They represent countries from six continents: Africa, Australia, Asia, Europe and Oceania, South and North America. This global, web-based education model yields innovative ideas for comparing urban and rural practice, and emphasises the importance of virtual learning processing, community participation, and exchange of service implementation strategies that are culturally relevant.

In Chapter 10, Lena Dominelli has the final word in this volume, being one of the most important representatives of the paradigmatic step that social policy and social work now are taking.[4] In her contribution *Environmental Justice at the Heart of Social Work Practice: Greening the Profession*, she argues that social workers have particular roles to play in assisting the mobilisation of people and resources when they wish to declare that environmental degradation is unacceptable and aim to formulate alternative models of socio-economic and environmental development rooted in the principles of environmental justice and a healthy, sustainable existence.

4 Her speech was delivered as the Katherine Kendall Memorial Award 2012 lecture at the Stockholm Conference. A comparable version is also published as an Invited Article in the *Int J of Soc Welfare 2013*:22:4.

Chapter 2

The Government of Canada: On Trial for the Racial Discrimination of First Nations Children

Cindy Blackstock

On 25 February 2013, the Canadian Government appeared before the Canadian Human Rights Tribunal to face allegations that its long-standing, flawed and inequitable provision of child welfare services to First Nations (Indigenous) children and families on reserves is discriminatory. The Tribunal has the authority to make a legally binding determination of discrimination and order a binding remedy enforceable in Canadian courts. These hearings mark the first time that the Canadian government is being held accountable for its contemporary treatment of First Nations children before a body capable of making enforceable orders (Blackstock 2011). This contribution describes the human rights complaint brought by First Nations against the Government of Canada and explores its implications for addressing other inequities experienced by Indigenous children in Canada and other countries.

The Government of Canada has a long and tragic history of separating First Nations children from their families. Beginning in the 1870s, the government forcibly removed First Nations children aged 5 to 15 years of age and placed them in residential schools run by Christian churches and controlled by the federal government with the aim of assimilating First Nations peoples (Royal Commission on Aboriginal Peoples 1996, Milloy 1999). Children attending the schools were banned from speaking their Indigenous languages and practicing their cultures as well as their spiritual traditions (Royal Commission on Aboriginal Peoples 1996, Milloy 1999). Physical and sexual abuse and death from preventable causes of disease were rampant and although people of the period, including a well-known physician, would routinely report the maltreatment of First Nations children, the government routinely failed to take adequate measures to ensure the children's safety (Bryce 1922, Caldwell 1967, Royal Commission on Aboriginal Peoples 1996, Milloy 1999). Of particular concern was the government's failure to properly fund the school buildings and operations despite credible reports linking the inadequate funding to the neglect, maltreatment and deaths of children (Bryce 1922, Caldwell 1967, Royal Commission on Aboriginal Peoples 1996, Milloy 1999).

The last residential school closed in 1996 and the Government of Canada apologised to former residential school students in 2008 (Harper 2008). The apology, while a meaningful gesture, did little to fix the multi-generational trauma and it did not incentivise the Federal Government to remedy First Nations child welfare funding shortfalls that contribute to growing numbers of children in care (Blackstock 2011).

Tragically, there are more First Nations children in child welfare care today than at the height of residential schools (Blackstock 2003) and they are entering into the care of child welfare at 6–8 times the rate of other children (Auditor General of Canada 2008). Studies show that the over-representation is substantially accounted for by the deeper levels of First Nations poverty, inadequate housing and substance misuse (Trocmé et al. 2006, Auditor General of Canada 2008) that have been shaped by the historical disadvantage rooted in the residential school era. For example, First Nations students at residential school typically only received primary school education as the government viewed First Nations children as academically incapable of further progress and thus being only suited to domestic servitude or farm work (Milloy 1999). The systematic deprivation of proper education coupled with the trauma related to cultural dislocation, abuse and neglect spurred multi-generational cycles of trauma and poverty that continue to undermine First Nations families today (Blackstock 2011).

Beginning in the 1980s First Nations began developing First Nations child welfare agencies to deliver culturally relevant services intended to target the unique factors placing First Nations children at risk. While provincial child welfare laws apply equally on and off reserves in Canada, the federal government funds child welfare on-reserve whereas the provinces fund it for all other children. However, the federal government funds First Nations child welfare on reserves at less than 80 per cent of the funding levels provided to children off-reserve by provincial governments even though the needs of First Nations children are greater (McDonald and Ladd 2000, Loxley et al. 2005, Auditor General of Canada 2008, United Nations Committee on the Rights of the Child 2012). This inequity is most prominent with regard to the range of services intended to keep children safely in their homes. Thus the Government of Canada's own reports have linked both the inequitable funding levels and the dated structure of the funding regimes for First Nations child welfare as contributing to the growing numbers of children in care (Blackstock 2011). As the Auditor General of Canada (2008) found, the Government of Canada has known for many years that its funding for First Nations child welfare is flawed and inequitable and yet has failed to take sufficient measures to remedy the problem despite available solutions developed in partnership with First Nations and one of the wealthiest national treasuries in the world.

Concerned about the impacts the inequality for First Nations children and their families, the Assembly of First Nations, the representative body for First Nations in Canada, and the First Nations Child and Family Caring Society of Canada (Caring Society), a research, education and policy organisation serving First Nations child welfare agencies, filed a complaint pursuant to the Canadian Human Rights Act alleging that the Government of Canada's inequitable flawed structure of child welfare funding was discriminatory.

The complaint was filed in 2007, and during the next seven years, the Government of Canada spent over 3 million dollars in its unsuccessful efforts to derail a public hearing on the facts by challenging the jurisdiction of the Canadian Human Rights Tribunal to hear the case. Contrary to its commitments pursuant to the United Nations Convention on the Rights of the Child, none of Canada's unrelenting arguments to dismiss the case were based on the best interests of First Nations children. Rather, Canada argued that the case should be dismissed as federal funding should not be compared to provincial funding for child welfare even though the same laws apply; in the alternative, the Government argued that the case should be dismissed because the Government of Canada funds First Nations child welfare agencies that deliver the service so if the service is discriminatory, it is the agencies that should be held accountable not the government (Blackstock 2011). It also suggested that the Canadian Human Rights Tribunal was not the appropriate place for this case to be heard. Three separate judgments by the Federal Court (Blackstock 2011, Federal Court 2012) and one by the Federal Court of Appeal (2013) dismissed Canada's arguments and ordered the hearings at the Tribunal to proceed.

The historic nature of these hearings cannot be overstated. The Canadian Human Rights Commission, a group charged with upholding the Canadian Human Rights Act, suggests that this case is one of the most important human rights cases in Canadian history and holds the potential of setting a legal precedent that could force the Government of Canada to provide equitable services to First Nations children in education, health and other areas (Langtry 2011).

The complaint is historic in nature but only a handful of people attended the news conference announcing the filing of the complaint in 2007. Despite the disappointing interest, the Caring Society believed that the Canadian public would be outraged if they knew about the inequitable treatment of First Nations children and were provided with a meaningful way to help. In that spirit, the Caring Society launched an online campaign called "I am a witness" which complements the human rights complaint. The "I am a witness" website provides the court documents of all parties as well as relevant reports by independent groups such as the Auditor General of Canada. Citizens and organisations are then asked to register as a witness meaning that they agree to watch the case online or in person before making up their own minds about whether Canada is treating First Nations children fairly. To date, over 12,000 individuals and organisations across Canada and around the world have registered to be witnesses online and hundreds routinely attend the hearings, making the First Nations child welfare human rights case the most formally watched human rights case in Canadian history. The increased public engagement in the case spurred increased media attention and international interest.

The international implications of this case for the rights of Indigenous children are significant. The United Nations Committee on the Rights of the Child (2003, 2012) has repeatedly noted that the Government of Canada's funding for First Nations child welfare is insufficient and UNICEF (2013) noted the importance of the Canadian Human Rights Tribunal as a redress mechanism for discrimination experienced by First Nations children. As the United Nations Committee on the

Rights of the Child notes in its general comment on the rights of Indigenous children, Indigenous children frequently experience discrimination including the receipt of inequitable services that often do not account for the higher needs of Indigenous children rooted in historical disadvantage nor do these services respect the rights of Indigenous children to their culture, religion or languages. This includes Indigenous children in other wealthy developed countries such as the United States, and Australia where Indigenous children receive inequitable child welfare services and other supports. Leaders in Indigenous child rights from both countries have studied the Canadian Human Rights Tribunal on First Nations child welfare and travelled to Canada to attend at the hearings to see if a similar approach could end longstanding government based inequities for Indigenous children in their respective countries. During Canada's Universal Periodic Review before the United Nations Human Rights Council, the United States and Norway both noted Canada's inequitable First Nations child welfare funding and urged corrective reform (United Nations Human Rights Council 2013).

The Canadian Human Rights Tribunal on First Nations child welfare is a useful example of how Indigenous peoples are employing binding domestic and international human rights mechanisms and leveraging public education and engagement to rectify longstanding Indigenous child rights violations. The Tribunal is expected to hear evidence through to the summer of 2013 with a ruling expected in late 2013 or early 2014. While First Nations are confident that the Tribunal, upon hearing all of the evidence, will conclude discrimination has occurred, there is no doubt that this case will set an important national and international precedent on Indigenous child rights.

References

Blackstock, C. (2003) First Nations child and family services: Restoring peace and harmony in First Nations communities, in *Child Welfare: Connecting Research Policy and Practice*, edited by K. Kufeldt and B. McKenzie. Waterloo, ON: Wilfred Laurier University Press, 331–342.

Blackstock, C. (2008) Reconciliation means not saying sorry twice: Lessons from child welfare in Canada. *From truth to reconciliation: Transforming the legacy of residential schools.* Ottawa: Aboriginal Healing Foundation.

Blackstock, C. (2011) The Canadian Human Rights Tribunal on First Nations Child Welfare: Why if Canada Wins, Equality and Justice Lose. *Children and Youth Services Review*, 33, 187–194.

Bryce, P.H. (1922) *The story of a national crime: an appeal for justice to the Indians of Canada.* Ottawa: James, Hope & Sons.

Caldwell, G. (1967) *Indian residential schools: A research study of the child care programs of nine residential schools in Saskatchewan.* Ottawa: The Canadian Welfare Council.

Federal Court. (2012) *Ruling: Canadian Human Rights Commission, Attorney General of Canada, First Nations Child and Family Caring Society of Canada, Assembly of First Nations, Chiefs of Ontario and Amnesty International Canada. Docket Numbers: T-578–11, T-630–4 and T-638–11; Citation: 2012 FC 445*. Toronto: Federal Court of Canada.

Federal Court of Appeal. (2013) *Ruling: The Attorney General of Canada v. Canadian Human Rights Commission, First Nations Child and Family Caring Society, the Assembly of First Nations, Chiefs of Ontario, Amnesty International and the intervener the Canadian Civil Liberties Association.* Docket Number: A 145:12. Ottawa: Federal Court of Appeal.

First Nations Child and Family Caring Society of Canada. (2013) *I am a witness*. Retrieved 6 January 2013 from www.fnwitness.ca.

Langtry, D. (2011) Speaking notes for David Langtry, Acting Chief Commissioner of the Canadian Human Rights Commission – At a News Conference about Extending Full Human Rights Protection to First Nations. Ottawa: National Press Theatre. Retrieved April 3, 2013 from http://www.chrc-ccdp.ca/media_room/speeches-eng.aspx?id=655.

Loxley, J., De Riviere, L., Prakash, T., Blackstock, C., Wien, F., and Thomas Prokop, S. (2005) *Wen: de: the Journey Continues*. Ottawa: First Nations Child and Family Caring Society of Canada.

MacDonald, R. and Ladd, P. (2000) *Joint National Policy Review of First Nations Child and Family Services*. Ottawa: Department of Indian and Northern Affairs Canada and the Assembly of First Nations.

Prime Minister of Canada Stephen Harper. (2008) Prime Minister Harper offers full apology on behalf of Canadians for the Indian Residential Schools system. Retrieved March 19, 2013 from http://www.pm.gc.ca/eng/media.asp?id=2149.

Royal Commission on Aboriginal Peoples. (1996) *Report of the Royal Commission on Aboriginal Peoples.* Ottawa, ON: Indian and Northern Affairs Canada.

Trocmé, N., MacLaurin, B., Fallon, B., Knoke, D., Pitman, L., and McCormack, M. (2006) *Mesnnmimk Wasatek: catching a drop of light: Understanding the over-representation of First Nations children in Canada's child welfare system: An analysis of the Canadian Incidence Study of Reported Child Abuse and Neglect (CIS-2003).* Ottawa: First Nations Child and Family Caring Society of Canada.

United Nations Committee on the Rights of the Child. (2003) *Concluding Observations: Canada*. Geneva: United Nations Committee on the Rights of the Child.

United Nations Committee on the Rights of the Child. (2012) *Concluding Observations: Canada*. Geneva: United Nations Committee on the Rights of the Child.

United Nations Human Rights Council. (2013) *National report submitted in accordance with paragraph 5 of the annex to Human Rights Council Resolution 16/21: Canada.* Working Group on the Universal Periodic Review, Sixteenth session. Geneva: United Nations General Assembly.

UNICEF. (2013) *UNICEF supports equitable treatment of First Nations children in on-reserve child welfare funding*. Retrieved March 13, 2013 from http://www.unicef.ca/en/blog/unicef-canada-supports-equitable-treatment-of-canada's-first-nations-children-in-on-reserve-chi.

Chapter 3
Environmental Social Work: A Future of Curiosity, Contemplation and Connection

Fred H. Besthorn

I believe it is becoming increasingly important to consider the current state of knowledge and future prospects in social work's efforts to become an active player in developing solutions in the context of global environmental crises.

Introduction

First, I want to say I am not an expert on this topic. Really, I don't view myself as an expert on anything. Not on life, not on teaching, not even on social work, and certainly not on the environment. But, one thing I have learned in the 60 years that I have been privileged to live on this exquisitely green, brilliantly textured, blue watered orb located in the far outer reaches of the spiralling Milky Way Galaxy; it is that we need far fewer experts and far more visionaries and doers of the hard work of environmental action. It is truly in the lives of those ordinary people who live in purlieus of Mumbai, the large *favelas* like Rocinha of Rio de Janeiro, the *barrios* of La Paz or the dusty, rural and impoverished village outposts of Malawi where the lasting, most meaningful solutions to the environmental problems that plague our world will come. It is in the lives, views and dedication of everyday people which come the most profound insights into creating a world fit for all.

This certainly was the case for the small community of Greensburg located in my home state of Kansas. Greensburg had a population of around 2,000 residences and was considered a thriving, historic and beautiful oasis in the semi-arid regions of southwest Kansas. It had everything a small community was proud of: a vibrant downtown area, its own small businesses, several community schools and all the accoutrements common to a small rural farming community. It was a community with many tree-lined streets, small park areas, and quaint and historic older homes. It was a place where everyone knew everyone and which took pride in its small town friendliness and down home, can-do spirit. Then came the late evening hours of 4 May 2007! It was a night no one would ever forget, in fact could never forget, and a night which changed the lives and consciousness of many hundreds of people and altered the course of the community forever.

On that evening of 4 May 2007 Greensburg, Kansas was hit by an EF-5 (Fujita Scale) tornado – the highest category of tornado severity. This storm was almost 2.2

kilometres wide with estimated sustained wind speeds of nearly 500 km/h. At the time, it was thought to be one of the largest, most destructive and most powerful tornados ever to touch down in the continental US. Few, if any structures, are able to withstand the devastating cyclonic winds of an EF-5 tornado. Some described it as if a giant hand had just swiped across the town, obliterating everything in sight. Others likened it to the aftermath of a nuclear blast. Whatever the metaphor of destruction, in the end Greensburg was literally wiped off the face of the earth, 95 per cent was utterly destroyed.

Kansas in located in what is notorious referred to as "tornado alley" – a swath of the central plains region of the US which experiences hundreds and in some years thousands of tornadoes each spring. For those of us who live there, we are used to wailing sirens warning of on an impending tornado, of the ominously darkening and blue/green wall cloud formations which might be the harbinger of death and destruction. But, the magnitude and seeming maleficence of this storm shocked even the most hardened of those who have lived their lives under the threat of killer storms.

This event might well have spelled the demise of Greensburg, Kansas. In fact, many said that what remained of the town should be levelled, residents relocated to other nearby towns and cities and the area of the town returned to farmland. The task of rebuilding was just too overwhelming, too many memories of that dark night in 2007 loomed, and no one knew where to begin.

Rebuilding a Community, Restoring Hope

However, the story of Greensburg is not just a story of tornadoes, tragedy and terror. It is not even a speculative story of changing climatic events, more severe storms, perhaps the by-product of global warming and unpredictable changes in the weather – although, in fact, many made the case for that argument. The destruction of Greensburg became a story of resilience and, most importantly, of vision.

In the five years following the Greensburg tornado the town did in fact rebuild. That might be expected but what was not expected was the way this small town decided to live again. The people of Western Kansas in the very heart of the United States are very common, ordinary, hardworking people. They are fiercely independent, generally politically conservative, and most don't particularly like "tree huggers" like me. But these common, ordinary people decided to do something quite extraordinary. After six months of spirited and sometimes contentious town meetings – participatory democracy exhibited at its best, the town – all the town, not just the surviving leadership, decided to rebuild Greensburg in a manner which was befitting its name. It would be rebuilt as a "Green" "Burg". In the five years that have followed the 2007 tornado, Greensburg, Kansas became the first "all green" city or town in the United States. The tragedy turned into a spectacular opportunity to do something that had never been done before. It was driven by the insights and visions of people who knew their town best. No experts, no centralized planning commissions, no utopian dreamers had told them what to do – though many had tried. What happened to Greensburg in the aftermath of the 2007 tornado was led first and foremost by everyday people who thought that a new life could and must be a "new" life patterned on ecological sustainability, participatory democracy and a deep seated willingness remake a community that not only served the residents but was able to sustain the environment that sustained them.

The homes which dot the landscape of Greensburg today are built with recycled materials, solar- or wind-powered, and represent the leading edge of sustainable environmental design. Solar power makes up a great deal of the energy needs of this new Green Burg. Green space is priority for Greensburg, ample walking and biking paths and restrictions on the use of automobiles in certain areas. In short, Greensburg found a way back and a way ahead. The movement was not lead by experts, although there certainly was expert support and advice along the way. Greensburg takes pride in having received the Platinum certification from the prestigious Leadership in Energy and Environmental Design (LEED) group. It residents now see the irony of what happened that night in 2007. There was great curiosity around what could be a new story for Greensburg. The unanticipated and rare opportunity to start over from the very beginning was an unexpected blessing in what otherwise would have been a dismal ending to a sad story. The vision, the hope, the planning, the cooperation in the aftermath of the tornado in Greensburg has some lessons for social work as well.

Environmental Awareness in Social Work

The often tepid and at times lack-lustre interest and performance of social work in the context of environment crisis has begun to change dramatically in the last decade. The profession was initially reluctant to engage in environmental issues and the environmental movement – content to restrict its environmental awareness to social context or individual development. Prior to the late 1980s and early 1990s little could be found in the academic or practice literature concerning the impact of deteriorating ecological condition and their influence on our clients and the larger world community.

This year marks the 20th anniversary of certainly not the first but arguably one of the most widely publicised and eagerly anticipated gatherings of the world community to address pressing environmental concerns. The Rio conference of 1992 in Rio de Janeiro Brazil was a hopeful gathering and marked by the emergence of a new international consensus that environmental issues must become a top, worldwide priority.

And it was about this same time 20 years ago that social work began to take its first tentative steps toward engagement in environmental concerns. Certainly, social workers have known for many previous decades that the environment is important to understanding the complexity of the human condition. But, partly due to the profession's own restricted theoretical models and their embeddedness in a highly individualised, economistic, and growth obsessed knowledge system characteristic of modern, mainstream Western worldviews, the profession was satisfied to see the natural environment as backdrop or content to our major concerns and, in the main, the responsibility of other professions.

But, as mentioned, in previous years the flowering of the profession's interest and concerns for issues of the natural or physical environment has been almost breath taking, especially for many of us who have been involved for quite a number of years in what some have come to call "environmental social work". John Coates and Mel Gray (2011) in a recent issue of the *International Journal of Social Welfare* review the history – particularly in North America and Australia – and the many, varied interest areas and themes with respect to the environment that have become a part of the growing body of professional knowledge. And, in the same issue, Jeff Peeters (2011) does a masterful job tracing the outlines of an emerging environmental social work in the European community. My own search of the literature has identified over 300 publications by social work scholars, practitioners, and students who have written on some dimension of social work with respect to the natural environment (Global Alliance for a Deep Ecological Social Work, 2012). Coates and Gray (2011) also note that there are likely many scores of other professionals who are doing environmental social work without realising it or without necessarily calling it by name.

For me these are encouraging signs and foreshadow positive trending for the years ahead. The question then becomes what do we as a profession need to sustain and strengthen this evolution? What values are important for the profession as we move ahead? I had been thinking on these questions for quite some time and frankly was at a bit of loss.

Then, several months ago my then 15-year-old daughter was checking out You-tube videos when she excitedly called out for me to see a clip that she had stumbled upon. She thought it was "cute" and, of course, when you are 15 and growing up in the west, cute is an important dimension of life, although I still have to admit I still haven't quite figured out what "cute" really means. But, I had to admit as I watched the short clip I also thought it was cute – even if I couldn't define what cute was.

The video was called "Jazz for Cows". In the clip, a jazz group begins playing for a herd of cows. The cows are mesmerised with the music, coming closer to the fence and growing in numbers as the band finishes their serenade.

Upon reflection the video became more than cute, it was something far more intriguing and interesting. I had always been around domesticated animals – cattle, horses, goats, and others and I was always fascinated by their ways. I had always felt from my earliest years that if we paid attention to the natural world and all those non-human beings around us they have something profound to teach us. Domesticated cattle are certainly no exception.

Curiosity

As I watched this short video, suddenly it dawned on me that these, our domesticated friends, had something to teach us. And, I think they may have something to teach our profession as we consider the future of environmental social work.

First, these bovine beauties were intensely CURIOUS. Our domesticated friends possessed an intense curiosity about the world around them. Indeed, all animals seem to inherently possess an innate curiosity. At some level, curiosity implies questioning the status quo – does it not? It means wondering if accepted knowledge, the standard and mainstream way of doing things is, indeed, our best place to begin. Curiosity in many ways is the wellspring of individual, collective and institutional genius.

Curiosity also means a recognition that the world in full of variety, possibility, and that there is no single, correct way to either understand the world or craft solution for issues and problems. In a real sense, curiosity is, as one author put it, 'an unwillingness to accept *yes* for an answer'. Sometimes we become too accustomed to saying "yes" to what is. We are told what to believe, experts pledge to us its veracity, and institutions demand that we cooperate. Corporations tell us they know what they are doing, and we are assured that the current path is the only path. In the United States we have a saying that captures this; we say "go along to get along". That means just accept what is and there will be no problems. Now I am not advocating that we be nasty and brutish in our questioning of what is. God only knows how much we are in need of much greater civility in an era when rudeness and offensiveness seems to have become social goods.

But, perhaps social work needs to rediscover its own sense of curiosity. Is it really possible that social work has something meaningful to contribute to the international crisis of the environment? Are we curious enough to find the unique, novel, unorthodox ways to contribute to the discussion? Curiosity and the diligent questions that arise from it reflect our core purposes in life and influence the very quality of what and who we are. It is likely that anything that you have ever done was first and foremost driven and directed by curiosity. The early contributors to environmental social work were curious practitioners and scholars. They wondered why, for instance, if 'social work trumpeted so loudly its person-in-environment perspective' how it could be that the *environment* – the very physical and natural world which sustains – us could have somehow been left out of our understanding.

The overriding question that currently drives the curiosity of many environmental social workers is "where does the profession go from here?" Let us be curious, let us not accept "yes" as the answer, let us allow our imagination to run wild and free for a time. Let us ask the questions that we know we have, as of yet, no answers for. Let us learn from our domesticated friends that the authority-pleasing, question-suppressing, rule-following approach that we all have been indoctrinated into may provide a society with a patina of security but does very little to prepare us for a new world, a new way of being, a deeper understanding and sense of interrelatedness to all that is around us. One of the problems of our profession, as with many others, may not be that "we don't have good answers" but rather "we don't have good questions".

Contemplation

A second thing we can observe from our domesticated friends is that of CONTEMPLATION. In the video, the cattle appeared to be standing in rapt attention as the jazz band played. It appeared that they were seriously and quietly contemplating what they were seeing. Of course, I don't pretend to know what a cow thinks. But, I am persuaded that all animals are thinking, feeling, communicative and contemplative beings. They pay attention! They unhurriedly study their world, they take time to process what is around them, and they do it quietly and intensely. They are not driven by time, schedules, and deadlines.

Unfortunately, the unhurried contemplative life is lost to most of us today. We live in a world of fast and furious sound bites. Quick food, quick glory, and quick answers are the mainstay of most of our existence in the western world. The absence of silence has become a disease in western culture. Our world is constantly, incessantly and unceasingly noisy. So much so that we often fear silence. The world is not only noisy it is a highly verbalised world where we too often defer to "experts" to tell us what to think, what to believe and how to respond. Taking time to be silent and contemplate is the beginning of deeper understanding. Ask yourself "where are you when you get your best ideas?" Most often people will say while resting in bed, walking in nature, listening to gentle music, gazing serenely at the stars or passing cloud formations. When you are alone and totally silent you are most completely yourself.. And when you are most completely yourself you are best able to craft the genuine questions and perhaps even find the meaningful solutions that changes life.

I have become convinced over the years that social work sometimes fails at its appointed tasks because we "work too hard", "play too little", "rest too infrequently" and are "noisy too much". It is essential that we take time to contemplate where we are and where to go from here. The best time for contemplation often comes when we are in an extraordinary, unaccustomed place. We cannot define our world quite as easily as we do our normal settings so we are left just to sense it, to see it and to contemplate it. That's good. That's what the residents of Greensburg had done. Their normal world was gone.

My daughter keeps telling me that I need, in fact must have, an electronically sophisticated and digitally vivid iPhone. I have politely reminded her that my current phone works just fine and so far I have been able to keep her prodding at bay. She asks why, saying a new phone would help me to better keep in touch. My answer is always: 'but sometimes I want to be out of touch more than I want to be in touch'. To be out of touch does not mean to be uninvolved. It means to be involved in a manner which is fully present, attentive to the new, the novel, and the unexpected not always encumbered by the white noise of distraction. Perhaps as we contemplate our profession's continued role in environment concerns we need to take a little more time to be "out of touch" so that we can do the quiet, contemplative work of determining where we go from here.

Connection

Finally, our domesticated friends teach us something of CONNECTEDNESS. As observed in the video, the cows gathered to attend to the music coming from these strange beings carrying these outlandish instruments just across the fence line. They gathered – together. There connection was not forced. It was natural, spontaneous and considerate. There was no pushing or shoving, no frantic attempt to get to the front of the line, no effort to stand-out that others might pay attention to the individual.

Social work knows, better than most, the critical importance of connection and relationship. We have been schooled to understand that meaningful life, important decisions, indeed the very essence of survival comes in the context of relationship with *the other*. Our profession's widespread adoption of contemporary systems theory as a guiding theoretical framework for the way we practice social work has taught us how everything connects to everything else. The secret to unparalleled creativity comes as we gather together to practice the fine art of combining and connecting disparate elements and ideas to form new patterns, new perspectives and innovative ideas.

And yet, it sometimes seems that we in the human species are less connected than ever before. Is it not ironic that we have electronic and institutional ways to connect with others in manners never before known in human history? And yet, we seem more divided than ever before. We seem to suffer from a heavy preoccupation with determining what separates us rather than with what unites us. Difference, division and divide in contemporary culture are good. Unity, utility and union are fearful and to be avoided. We are, more often than not, driven by our differences and divisions over religion, cultural, ethnicity, race, political affiliation and sexual orientation. We sometimes seem to have become too satisfied with the patina of connection while the depth of collective joining eludes our grasp.

I recently had an experience at my university which starkly illustrated this for me. It was a busy day of classes, students and faculty were rushing from one place to another and about 10 of us converged on an elevator to be transported to the upper floor of our multi-floored classroom complex. I stepped on board the elevator with about 10 students. As we all rode to the tops floors I could not help but notice that every single student had their eyes buried in and their attention occupied with cell phones – some listening to music with ear buds tucked tightly into their ear canals, others texting feverishly as if some world catastrophe had just been announced, and others scanning robotically through page after page of some unknown website. In the few minutes that it took to deposit each of us at our appointed destination not a word was spoken, not a glance was exchanged, not a courtesy was offered. I felt out of place. I had no phone, I had hoped for a bit of playful banter, perhaps for an opportunity to say good morning or to exchange a greeting. But there was none of that. We stood alone in our 3x3 foot square of personal space – all in our own individual worlds, occupied by electronic diversion – almost afraid to find a point of meeting. Indeed, we were together in the same space but we were not in any manner connected.

Connectedness in Environmental Social Work

From my perspective, one of the things that environmental social work most desperately needs in the years ahead is to find connection. First, we need to connect to the rest of our professional colleagues who may not understand and/ or who fail to see the relevancy of the environment to the practice of social work. Obviously, many who read this may have had that experience. You may remember the awkward moment when you tried to explain to a professional colleague why environmental issues were important to you and to the profession. I remember the first years of my own practice and academic work. I would often get quizzical stares and comments like: 'Environmental what … ? You must be joking. You want me to worry about the environmental crisis when these social crises play themselves out in front of me each and every day?', 'Can you explain just one more time what you mean by this idea of "deep ecology?"', 'Wow, it must be so deep that no one understands it!'

Each day, I suspect hundreds and hundreds of professionals practice environmental social work – whether they completely recognise that or not. And each year there is growing recognition that unless our profession collectively attends seriously and persistently to a collapsing environment there will one day be no social work – indeed there may one day be no world at all.

Secondly, social work must find the way and the resources to make connections with other professionals who share our commitment and concern for environmental issues. Whether we call this work interdisciplinary, trans.-disciplinary or multidisciplinary the underpinning ideas are the same. Again, John Coates and Mel Gray (2011: 2) have pointedly captured this important need. They note that 'the challenge today is for social work and other professions to break free of their academic silos and work cooperatively or, at least in union, to better understand and respond to the many dimensions of climate change and environmental destruction'. Yes, I could not agree more.

Our work has been historically an isolated endeavour. One of my students once said, reflecting on the history of social work in the US: 'we don't seem to play well together with others'. Of course others might say that 'others do seem to play well together with us'. Both are undoubtedly true. But the end results are the same.

Finally, harkening back to what I said earlier, our profession must find ways to connect to the home-grown, grassroots, unadorned efforts of everyday persons and communities who are doing the best and most lasting environmental work. Those most impacted by deteriorating environmental conditions are often the most resourceful and insightful. The poor, marginalised and ecologically disadvantaged must not and cannot wait passively for the west to save them. And, fortunately, they are not. We in the west have been historically fond of grand, utopian ideas that, while sometimes well-meaning, have too often been shrouded in the incipient arrogance of cultural superiority and technocratic hubris. As William Easterly, former World Bank economist notes, we in the west are awash in bureaucratic experts while the rest of the world works unobtrusively utilising active, grassroots civic involvement. He calls it the difference between the *planners* and the *searchers*. The expert planners embrace the *big plan* funnelled through large, governmental schemes. The searchers are everyday people who seek out small, doable actions which improve and change the lives of those most impacted. Easterly notes, 'the big plans are attractive to politicians, celebrities, and activist who want to make a splash, without the Western public realising that those plans at the top are not connected to reality at the bottom' (Easterly 2006: 17).

Conclusion

Social work's contribution to an ecologically just and sustainability world system is to be about what it has always done so well. It is about joining with those local groups who engage in sometimes piecemeal, sometimes intermittent, even sometimes ineffective political and social action to relieve suffering, and work to sustain natural environments which don't get bogged down in big plans with infeasible objectives. If we ask the question 'what can social work do to eliminate environmental crisis?' we are likely asking the wrong question. The more appropriate question is 'what can social work do to improve the lives of people?'; that is 'what can social work do to join with those peoples at local levels most impacted by environmental decline and most knowledgeable and prepared to take actions necessary to improve their unique situations?'

So, where have we come? As I said earlier, I am no expert. But, I think I do know a little about social work, its history, and its many, many efforts and important accomplishments over the decades. And this I do know; challenges are real, opportunities are abundant and transformation is possible. The world community and natural systems cannot wait. We are at or near the tipping point with respect to global environmental conditions and there is precious little time for us to take on our role in this international undertaking. It must begin with our genuine curiosity for what is possible, must be imbued with a spirit of contemplation, considered thought and deliberation concerning the next steps and must be sustained by a commitment to connection, bilateral and multilateral cooperation, and inclusivity of the broadest possible range of voices and perspectives. And finally it must fully incorporate the experiences, insights and real-world solutions that every day, hard-working people have into what is important and necessary to their well-being – in their place and at their time. This is what social work does, it is what we can do well and it is what we must do now.

References

Coates, J., and Gray, M. (2011) The environment and social work: An overview and introduction. *International Journal of Social Welfare, 21*(4), 1–9.

Easterly, W. (2006) *The white man's burden: Why the west's efforts to aid the rest have done so much ill and so little good.* New York: Penguin Books.

Global Alliance for a Deep Ecological Social Work. (2012) *Social work and environment bibliography.* Retrieved from http://www.ecosocialwork.org/indes.php?option=comcontent&view=aarticle&id=59&It mid=66.

Peeters, J. (2011) The place of social work in sustainable development: Towards ecosocial practice. *International Journal of Social Welfare, 21*(4), 1–12.

Chapter 4

Disaster Management: Perspectives for Social Work and Social Development

This is relatively a new field of action for social work. During the last years interest has strongly increased among professional social workers in being part of the responsible actors in prevention of disasters and in acute disaster management. The need for psychosocial support has been highlighted by the UN Office for the Coordination of Humanitarian Affairs, OCHA.[1]

This kind of support is very appropriate for professional social workers and training courses are being offered in some institutions.[2] The necessity of a holistic view as well as a gender perspective of people in natural disasters is creating a basis for inventions of crisis management, networking and psychosocial solutions, suitable for the social work profession.

Disaster Management – Action and Impact: Perspectives for Social Work and Social Development

Roland Sewell

Overview and Background

This contribution seeks to demonstrate the key role for social workers in disaster management. Social workers have much to contribute, their ethos is central to good disaster management; social workers can help minimise the negative impact of disasters and poor development but also optimise the positive links and opportunities from disasters to good development; they have a primary role at every stage of the disaster management cycle.

I have combined three rewarding and complementary vocations in one life time, but never within academia nor the social work profession, so I need to justify on what authority had I made these assertions to social workers at the 2012 Joint World Conference on Social Work and Social Development.

1 Follow the development on www.irinnews.org/Report.aspx?ReportID=90972rLink.

2 For example by Council on Social Work Education CSWE – Disaster Management.

The Institution of Civil Engineers records the definition of civil engineering by Thomas Tredgold for its royal charter in 1828. These words (ICE 1870: 215 note 1) continue to inspire me:

> Civil engineering is the art of directing the great sources of power in nature for the use and convenience of mankind.

Contrary to a common view, this illustrates that a civil engineer needs to be people-focused, even more so when those powers of nature get out of control! A minister of religion is authorised to perform certain religious ceremonies but his or her primary responsibility is the 'flock' especially in times of emergency. A relief worker seeks to alleviate suffering and the impact of disasters with human dignity. I therefore find much in common with social workers and feel justified and greatly privileged to be given a chance to share these perspectives on the topic of social work and social development within the context of disaster management and impact.

Twenty five years ago I was in conversation with a young female social worker who, with a little overseas experience, was leading a large team of relief workers in a massive response to an international emergency in Sudan and Ethiopia. The writer had been busily erecting water treatment plant and digging latrines around the camp, while she had been in conference with community representatives, the United Nations High Commissioner for Refugees (UNHCR), and relief agencies. I sincerely commended her for the important leadership role she was fulfilling. She may have thought me patronising or cynical; perhaps I unintentionally was for she rebuffed me, saying: 'The trouble with you engineers is that you are so busy with things that you don't find time to talk to people'. Chastened by this I have taken her words to heart and have sought, ever since, to emulate good social worker principles in what I do and more particularly, how I do it. Listening to people makes all the difference!

A Few Definitions

For the sake of this contribution the UNISDR 2009 web definition of a *Disaster* is used:

> A serious disruption of the functioning of a community or a society involving widespread human, material, economic or environmental losses and impacts, which exceeds the ability of the affected community or society to cope using its own resources.

Disaster Management applies to many disciplines but again the UNISDR 2009 web definition of Disaster Risk Management is adopted:

> … the systematic process of using administrative directives, organizations, and operational skills and capacities to implement strategies, policies and improved coping capacities in order to lessen the adverse impacts of hazards and the possibility of disaster.

These two definitions are clear-cut, but finding a concise definition of *Social Development* has been less easy. The links between social development and other development is critical but for the sake of this paper the opening paragraph from the Indices of Social Development (ISD 2011) website seems most fitting:

> Social development is about putting people at the centre of development. This means a commitment that development processes need to benefit people, particularly but not only the poor, but also a recognition that people, and the way they interact in groups and society, and the norms that facilitates such interaction, shape development processes

The Ethos of Social Work is Central to Good Disaster Management

In preparing for the conference, it came as a surprise to discover that the involvement of social workers in disaster management was perceived to be relatively new. Is that true? Social workers may not always have been evident in relief teams, but the ethos, values, language and principles of social work are central in shaping the policies, procedures and practices of good disaster management.

The Ethiopian Famine of 1984–86 was a defining moment in the international professionalisation of relief work. Drought in the Horn of Africa is a recurring threat but it was drought combined with war that led to a famine which affected close to 8 million people of whom over 1 million died. In 1984, a BBC news crew documented the disaster, with the journalist-news reader, Michael Buerk (1984) describing it on film as: ‘a biblical famine in the 20th Century’ and ‘the closest thing to hell on Earth’.

The report shocked Britain, stimulated world attention and brought about an international response from Britain, Sweden, West Germany, Poland, Canada, USA the Soviet Union and other nations. An international response is commonplace today but not then and coordination, cooperation, leadership and a common set of operational standards were sadly lacking.

Back then, most relief workers were short-term volunteers and personnel seconded from Europe, with little experience of the social needs of the community they were working with. Thankfully, building on lessons learnt from this and subsequent emergencies the requirement today is that relief workers have an understanding of the humanitarian and social context of their work.

The International Committee of the Red Cross (ICRC) and the Federation of Red Cross and Red Crescent Societies (ICRC and IFRC 1994) prepared a Code of

Conduct which international humanitarian organisations are expected to sign up to. Ten years later, Biber (2004) reported on the ICRC web site:

> The Code of Conduct is the expression of a common operational approach for providing help to those in need, based on strongly cherished principles and International Humanitarian Law.

Organisations are required to comply with the Humanitarian Code of Practice but they must also be fully embedded as fundamental values in the people in disaster management. What are the values of humanitarian workers?

Walker and Russ (2010: 29–38), writing on the research undertaken by ELRHA, a collaborative network dedicated to Enhancing Learning and Research for Humanitarian Assistance through a partnership with higher education institutions and humanitarian organisations; identified the top four values required of humanitarian relief workers as:

- *respect* for the beneficiary and their community
- *accountability* to beneficiaries and their community
- *independence* from political, financial, religious or other pressures
- *impartiality* of action.

From these a framework has been developed for those working in the humanitarian sector following a 2-year global consultation process led by the Consortium of British Humanitarian Agencies and People in Aid (Rutter, et al 2011: 34–38). This competency framework was developed to devise appropriate training courses for humanitarian workers. They are summarised thus:

- Understanding the humanitarian context
- Applying humanitarian standards and principles
- Ensuring programme quality and impact
- Accountability
- Decision making
- Listening and creating dialogue
- Working with others
- Minimising risk to communities, partners and stakeholders
- Managing personal safety and security
- Adapting and coping
- Maintaining professionalism
- Self-awareness
- Motivating and influencing others
- Critical judgement

Social workers reviewing the above disaster management competencies will recognise them as equally central to good social work practice.

Two years later Russ and Smith (2013: 32–36) reported on research undertaken by ELRHA using a questionnaire based on the CBHA framework and drew attention to the need to enhance training in the humanitarian sector. Interestingly, this revealed that although 74 per cent of respondents were happy with the CBHA list 26 per cent wished to add some or all of the following.

- 'Psycho-social/psycho-spiritual'
- 'Cultural sensitivity'
- 'Self-management'
- 'Motivation, communication, initiatives, and teamwork'.

Russ and Smith (2012): 32, single out the most frequently identified gaps in competency in the humanitarian sector as communicating and working effectively with disaster-affected populations, involving them in disaster response programmes and measuring overall performance against their expectations. Clearly humanitarian workers need to enhance social work skills.

The above not only demonstrates the commonality of values and competencies amongst social and humanitarian workers but illustrates a role and opportunity for social workers to contribute to improved training of humanitarian workers and in achieving good disaster management in the sector.

Social Work and the Disaster/Development Link

Although not always adequately appreciated, there is a growing awareness of the link between disasters and development. On the negative side, if mistakes are made in disaster management it can put at risk or undermine the recovery and future development of the impacted community and similarly, ill-conceived and badly implemented development can increase vulnerability to disaster. On the positive side, good development can reduce vulnerability and good disaster management can lead on to opportunities for development that may never otherwise have happened.

Stephenson (1994) illustrates and explains the relationship and vulnerability of development and disasters which for the purposes of this paper is also used to demonstrate the vital role of the social worker. In Fig 4.1, the disaster realm is represented below the horizontal axis and the development realm above. The positive realm is represented to the right and negative to the left of the vertical axis.

The Bottom Left Quadrant is where a Disaster has a Negative Impact upon Development

In this scenario, the disaster, (e.g. flood, earthquake, tsunami, or man-made) may destroy people, property and progress, divert resources (people, skills, money and time), undermine donor/investor confidence, de-motivate the community, divert strategic focus and, at best, hold up the development programme and, at worst, destroy it. Throughout the disaster response and recovery, the social worker has

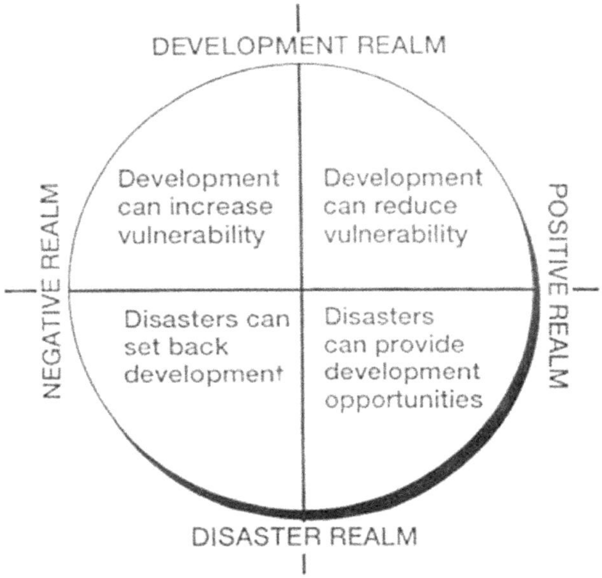

Figure 4.1 Development and Vulnerability to Disasters

Source: Stephenson 1994: http://www.untj.org/docs/Disaster_Management/Resources%20 Page/disaster_development.pdf

an important psycho-social, counselling, listening, safeguarding, facilitating and motivating function, and an important part to play in encouraging the restoration of hope and a future. Space does not permit to list examples but a little research will reveal them from such disasters as the 2011 Earthquake in Christchurch, NZ., the 2004 Tsunami in South Asia, or Hurricane Katrina in 2005 in New Orleans, US.

The Upper Left Quadrant is where Development can Increase a Community's Vulnerability to Disaster

Poorly-conceived and/or implemented projects such as urban development, industrial expansion, irrigation, water capture, power generation, transport, construction, forestry, agro-economic, coastal development or even tourism development can cause environmental and socio-economic threats such as overcrowding, mass movement of people, loss of productive lands and traditional livelihoods, reduction in staple food stuff production which may result in famine, disease, inadequate health provision or even civil unrest. The crucial role of the social worker is in recognising these threats, lobbying and mitigating their impact.

The Upper Right Quadrant Represents where Development Reduces Vulnerability to Disaster

Here, good development not only achieves real social-economic benefit but has the effect of reducing vulnerability to, preparing for or mitigating the impact of a disaster. Hazard-resistant construction or coastal and flood protection in vulnerable regions would be obvious examples but also the development of drought resistant crops or the improvement of coping mechanisms and services to vulnerable communities would come into this category. The skills of the social worker are key in collecting and contributing people centred information into the design process and in appropriately disseminating good practice in the implementation process.

The Lower Right Quadrant is where Disasters Open Up Development Opportunities

Where before a disaster there may have been no political will or financial resources for social, economic, or environmental development, a disastrous occurrence can promote a change for good. Media interest can raise international awareness causing a flow of resources, or a change in political priorities. Even where the disaster is not on a scale to promote international attention the local impact of the event can be the catalyst needed to bring about change. Social workers have a role in capitalising on such opportunities to draw communities closer; help stimulate community capacity building and empowerment; address vulnerabilities and encourage authorities to invest in hazard reduction and positive development.

Understanding these links between disasters and development helps to illustrate the value and importance of the social worker's role in disaster management.

Social Work and the Disaster Management Cycle

Figure 4.2 below is the author's illustration of the *Disaster Management Cycle*, which is a much-used and convenient model to describe the sequence and process by which governments, businesses, and civil society plan for; reduce the impact of; react during and immediately following and then recover from a disaster. Throughout this cycle social workers have a role to play. A brief search on the United Nations International Strategy for Disaster Reduction (UNISDR) website revealed 170 papers and references specifically to the various roles of social workers at all points of the Disaster Management Cycle.

Response ...

... may be a brief stage in the cycle or can last for days, months or years. Contingency plans kick in, rapid assessments are undertaken, emergency services respond and relief workers distribute aid. There is often a period of search and rescue but generally it starts by addressing the most fundamental physiological and safety needs (food, water, shelter, sanitation, health and security). These needs are essential to preserve life but as both Maslow (1954) and Manfred Max-Neef (1991) identified, humans have needs that go beyond the physiological. People's lives and livelihoods will have been affected, families separated, grief, loss and a sense of hopelessness will have found root. Good disaster management will ensure

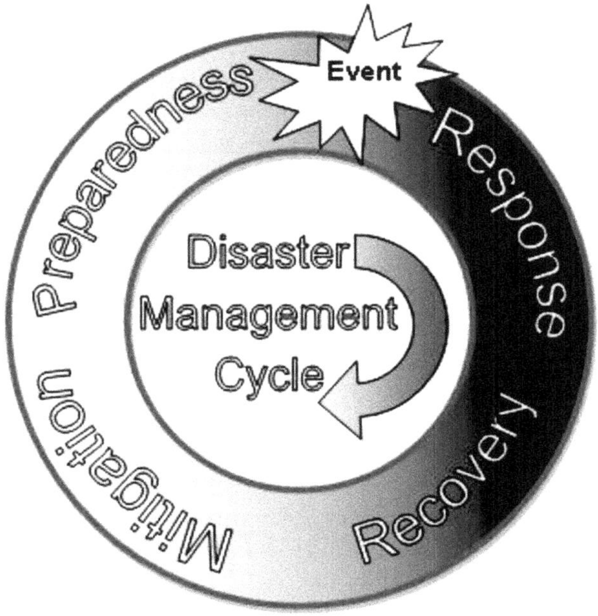

Figure 4.2 The Disaster Management Cycle (author's figure)

that at the earliest opportunity psychosocial support is available. The role for social workers is essential both to the victims of the disaster and for the responders.

Recovery ...
... is the stage of restoring all aspects of life that have been affected by the disaster's impact on a community. Social workers have great skills to offer here. In addition to invaluable psychosocial support post trauma, it is important that activities are established and maintained that re-build a sense of worth and community cohesion and a whole hearted involvement by the affected community in their own recovery. Regenerating enterprise is valuable in rebuilding hope and confidence with opportunities for businesses to move from temporary to permanent homes and regenerate employment and the economy. Social workers are needed alongside the community as the focus moves from emergency repairs to reconstruction and redevelopment.

Mitigation ...
... involves reducing vulnerability to the impact of disaster threats. Social workers are well placed to contribute here by highlighting the potential impact on individuals and most particularly on the more vulnerable in society who are often overlooked. An awareness and sensitivity to these needs and how they can be mitigated is often better appreciated by social workers as it often requires changes in local habit and behaviour as well as building codes and bye laws. Developing emergency

evacuation plans and other activities to reduce vulnerability and increase resilience benefit from the input of social workers who know the community and the most vulnerable within it.

Preparedness ...
Requires the understanding of how a disaster may impact the community and how education, outreach and training can build capacity to respond to and recover from a disaster. Material preparations need to be made for an imminent or potential emergency but social workers have a special role in psycho-social preparedness, thinking through and advising on the likely impact on individuals and contingency planning, looking at adequate provision to the most vulnerable. These need to be developed in dialogue with community leaders and communicated appropriately and where there is time, training scenarios should test out contingency plans. Social workers are key to good preparedness.

Conclusion

The conclusions from this paper are simple and hopefully provide a stimulus to further discussion and greater and effective involvement by social workers in Disaster Management.

- The social work ethos, its language, values, techniques and core competencies are common and central to good Disaster Management.
- There is a link between the effect of Disasters on social development and vice versa. Social workers can help minimise the negative and optimise the positive impacts.
- Social Workers have a key role at every stage of the Disaster Management Cycle.

References

Biber, B. (2004) *The Code of Conduct: humanitarian principles in practice* [Online: International Committee of the Red Cross Resource Centre]. Available at: http://www.icrc.org/eng/resources/documents/misc/64zahh.htm [Accessed March 2012].

Buerk, M. (1984) *Extent of Ethiopia famine revealed* [Online video clip BBC News 22 October 2009] available at http://news.bbc.co.uk/1/hi/in_depth/8315248.stm [Accessed 15 February 2012].

ICE. (1870) *Minutes of Proceedings of the Institution of Civil Engineers.* London: The Institution ICRC and IFRC. (1994) *The Code of Conduct for the International Red Cross and Red Crescent Movement and Non-Governmental Organisations (NGOs) in Disaster Relief 1994* [Online: International

Committee of the Red Cross] Available at: http://www.icrc.org/eng/assets/files/publications/icrc-002–1067.pdf [Accessed February 2012].
ISD. (2011) *Defining Social Development* [Online: Indices of Social Development] Available at: http://www.indsocdev.org/home.html [Accessed 15 February 2012].
Max-Neef, M.A. (1991). *Human Scale Development Conception, Application And Further Reflection.* Apex Press. Available as pdf: http://www.max-neef.cl/download/Max-neef_Human_Scale_development.pdf.
Maslow, A. (1954) *Motivation and personality.* New York, NY: Harper.
Russ, C. and Smith, D. (2012) *A Global Survey on Humanitarian Professionalization* [Online: Enhancing Learning and Research for Humanitarian Assistance (ELRHA)]. Available on: http://www.elrha.org/uploads/Global%20Humanitarian%20Professionalisation%20Survey.pdf [Accessed March 2012].
Rutter, L., Swords, S., Price, M., and Emmens, B. (2011) *Working together for people in crisis – Core Humanitarian Competencies Guide* [Online: Consortium of British Humanitarian Agencies]. Available on: http://www.thecbha.org/media/website/file/CBHA_Core_Humanitarian_Competencis_Guide_Finalpdf.pdf [Accessed February 2012].
Stephenson, R.S. (1994) *Disaster and Development: 2nd Edition* [Online: Disaster Management Training Programme. UNDP] Available on: http://www.untj.org/docs/Disaster_Management/Resources%20Page/disaster_development.pdf.
UNISDR. (2009) *Terminology on Disaster Risk Reduction* [Online: United Nations Office for Disaster Risk Reduction]. Available on: http://www.unisdr.org/we/inform/terminology [Accessed 15 January 2012].
UNISDR. (2012) [Online: UNISDR Various social work related papers] Available on: http://www.unisdr.org/search?cx=014932112152556794315%3Aev9g2xr_5ni&cof=FORID%3A11&i.e.=UTF-8&q=social-workers&sa.x=15&sa.y=15 [Accessed March 2012].
Walker. P., and Russ, C. (2010) *Professionalising the Humanitarian Sector A scoping Study* [Online ELRHA] Available on: http://www.elrha.org/uploads/Professionalising_the_humanitarian_sector.pdf [Accessed: February 2012].

Roles and Functions of Social Workers in Wenchuan Earthquake Post-Disaster Intervention: Based on the Three Stage Model

Xiying Wang and Xiulan Zhang

Introduction

It has been over four years since the Wenchuan Earthquake hit Sichuan and other areas of southwest China. The post-earthquake suffering was immense – about 87,000 people were dead or missing, more than 374,000 people were injured, along with 50,000 villages and towns destroyed, and direct economic loss of $125 billion (United Nations Economic and Social Council, 2008). This was indeed a major catastrophe. The disaster created an overwhelming, complex situation replete with challenges for the unprepared Chinese social work profession.

Social workers in Western societies have a long history of dealing with the aftermath of natural disasters (Ku, Ip, and Xiong 2009). Social workers play an important role in the disaster management circle, from preparation to mitigation and recovery (Zakours 1996). Furthermore, they are most effective at managing the symptomatology of post-traumatic stress, delivering relief resources and services to vulnerable populations (Galambos 2005, Mitchell 1983, Van den Eynde and Veno 1999; Zakours 1996), formulating and implementing different service plans (Banerjee and Gillespie 1994, Dodds and Nuehring 1996), organising and advocating for the community (Pyles, 2007), rebuilding social relationships, and restoring social functions (Miller 2012). Social workers' participation after natural disasters has always been 'assumed rightfully automatic' (Ku, Ip, and Xiong 2009:148). However, this is not the case in China. Social workers in mainland China are less prepared in post-disaster intervention than their Western colleagues (Ku, Ip, and Xiong 2009). They have very limited training in crisis intervention, case management, and direct mental health interventions, with no experience in post-disaster interventions before the Wenchuan earthquake.

The Wenchuan earthquake was 'undoubtedly a monumental disaster' (Ku, Ip, and Xiong, 2009:148), but it was also a defining moment for the Chinese social work profession. It is valuable to document the first-hand experiences of social workers involved in the disaster management of the different stages of rescue, temporary relocation, and reconstruction, to inform not only how the profession has grown stronger, but also how it influenced the rebuilding and recovery of affected areas. Most of the materials presented in this article come from personal experiences, as well as individual interviews and focus groups with social workers in mainland China, Hong Kong, and Taiwan.

Three-Stage Model

We use a three-stage model as a time framework, dividing the post-disaster intervention into unique stages: 1) rescue; 2) temporary relocation; and 3) reconstruction. In different stages affected individuals, families, and communities have different needs and wants. In turn, social workers play different roles during each stage to provide services and promote recovery (Feng 2008).

Because of the large military and government mobilisation after the earthquake, the rescue phase after the Wenchuan earthquake only lasted about a week and ended when the Chinese government shifted its focus from rescuing victims to taking care of the survivors. The second phase began when people who lost their houses were temporarily relocated to unaffected communities, sport centres, or makeshift tent villages and ended when they moved to temporary resettlement areas, which lasted from the first week to three months. The third phase began when survivors moved from sport centres, tent villages, or other temporary accommodation areas to temporary resettlement areas and ended when they finally moved back to their permanent houses. Although experiences from other Asian earthquakes, such as the 921 Earthquake in Taiwan in 1999 and the Kobe Earthquake in Japan in 1995, suggested that the reconstruction phase could take over a decade, the Chinese government successfully moved most of the earthquake survivors back to their permanent homes in less than three years.

Although we applaud the remarkable achievements by the Chinese government in the rescue and resettlement efforts, we argue that the psychosocial reconstruction process (i.e., grieving, recovering a sense of hope and purpose, and connecting with others), takes much longer than rebuilding houses and repairing roads. Among the tough tasks of post-disaster community reconstruction, a major challenge for social workers was to help the communities draw on community strengths and foster resiliency. In the later sections, we will examine in detail how Chinese social workers successfully linked 'physical safety and survival needs to social emotional coping and community bonding' (Tan, forthcoming), thereby promoting individual, family, community and social recovery, along with capacity building.

Social Work Interventions in the Rescue Phase

During the rescue stage, as the arduous relief work unravelled, the People's Liberation Army (PLA), medical personnel, volunteers, and teachers were generally considered to be the four most respected groups. They diligently carried out their roles and efficiently accomplished impossible tasks in the quake zones in a selfless manner (Sim 2009). In contrast, social workers in mainland China were not mobilised. Because of their lack of confidence and experience mainland Chinese social workers were uncertain about their roles in the disaster intervention efforts. Their uncertainty was also deeply related to a simple fact: China originally lacked a comprehensive national policy for disaster intervention. The Chinese central government only included psychologists, rather than social workers, as providers of post-disaster mental health interventions. At that time the Ministry of Civil Affairs, overseeing all social work practices in China, hesitated to take initiative and send social workers to the disaster areas. In particular, they were unsure about what social workers could do differently than psychologists or other helping professionals.

Moreover, local governments in the disaster areas did not understand professional social work practice and its potential to contribute to the rescue, resettlement, and reconstruction processes. Most local government officials did not know the differences between volunteers and professional social workers. At that time, local governments were already overwhelmed by the influx of more than 50,000 volunteers into the disaster areas and it was natural for them to worry that added social workers in the disaster areas would burden already strained areas. As a result, local governments discouraged social workers from entering the disaster areas.

Social Work Interventions in the Temporary Relocation Phase

The inaction of social workers in mainland China immediately after the earthquake changed into scattered mobilisation during the temporary relocation phase. On 20 May 2008, eight days after the earthquake, the first mainland China social work team from Shanghai arrived at the disaster area for an initial 3-day needs assessment. On May 27, the Ministry of Civil Affairs sent a team of 12 social work educators and other experts into the earthquake area. On the same day, the Guangzhou government also sent a group of ten social workers into disaster areas (Wang 2008). Many other central governing bodies and NGOs outside the disaster areas followed the same practice, sending their needs assessment teams to different disaster areas. Soon after the needs assessments these central governing bodies and NGOs started sending their professional social workers into the disaster areas to support relief efforts. Most of them stayed in the disaster area for one to two weeks and helped the survivors in the temporary relocation areas and makeshift tent cities. Outside the disaster areas social workers also served survivors who were relocated to their cities. During this period social workers accumulated experience responding to the disaster and performed various roles discussed, in turn, below.

Social workers as coordinators of multidisciplinary teams

In addition to providing direct services to earthquake survivors, many social workers also worked as coordinators of multidisciplinary teams. In many temporary relocation areas social workers usually worked side-by-side with other professionals, often assuming team coordinator roles.

Social workers worked as coordinators among organisations

Through doing the community audit and making comprehensive lists of organisations providing services, the social workers could identify service gaps between local residents and community organisations, and also developed new services that were not provided by other organisations in the community. For example, the social workers in Mianzhu Sport Center (a temporary refuge for earthquake survivors) supported groups for mothers with infants and toddlers, women who lost their husbands, and parents who had child(ren) in senior high schools facing the college entrance examinations.

Social workers promoted fair distribution

In the second stage distribution of relief supplies, such as tents, water, food, clothes and other supplies, was the local government's daily job. However, since the distribution adopted a top-down approach, these practices led to perceptions of unfair treatment among survivors. However, social workers are used to working with communities and are very sensitive to local needs, which made them excellent partners to governments in their relief efforts. Social workers also channelled the needs of survivors to potential donors and helped donors understand the priorities of local communities.

Social workers as conflict mediators

Social workers were involved in mediating conflicts between grieving parents who lost their children in the earthquake and the local governments who the parents blamed for their children's death (because of collapsed school buildings). In Mianzhu County, when the conflict between angry parents and local government escalated, social workers, organised by the National Working Committee on Children and Women (NWCCW), stepped in to diffuse the potentially dangerous confrontation. They helped parents voice their demands for justice and tried to divert direct confrontation that might harm parents or governmental officials.

Social Work Intervention in the Reconstruction Phase

During the reconstruction phase both central and local governments in disaster areas were more enthusiastic about providing social work services for earthquake survivors. Social service centres were established in the temporary resettlement areas and the once-fragmented and ad hoc social work intervention slowly turned into more organised long-term commitments (Lum, Wang, and Danso 2010). Beginning in mid-June governments outside the disaster areas, collaborating with social work departments in colleges and NGOs, started to deploy social work teams into the disaster areas to establish social service centres.

By the end of 2009 more than 30 social service centres had been established, and more than 1,000 social workers entered earthquake-affected areas to provide all kinds of services within the first year of the earthquake (Liu 2009). At this stage one third of these centres were established through collaborations with NGOs and universities from Hong Kong and Taiwan. Besides the social workers from cities outside the disaster area, such as Guangzhou, Shanghai, and Shenzhen, the social work educators played an important role in the disaster relief and reconstruction process. During the third phase the roles of social workers expanded to include community organising and advocacy, community economic development, case management, rebuilding social relationships, restoring social functions, continuously providing disaster mental health interventions, grief counselling, and running support groups for various vulnerable populations in the temporary resettlement areas.

Firstly, social workers as facilitators of collective and individual grieving and mourning

Social workers helped earthquake survivors feel more positive, hopeful, and connected. Chow, Chan, and Ho (2007) argue that there are unique factors to the judgment of a "good death" in China: 1) knowing that the cause of death 2) was natural and timely and 3) the deceased did not suffer. However, the Wenchuan earthquake caused a lot of sudden, untimely, and violent deaths, resulting in painful and overwhelming grieving experiences for those who lost their loved ones. On the individual level social workers provided emotional support to bereaved people and organised self-help mutual support groups for those with similar experiences. On the collective level they worked with people to encourage culturally meaningful forms of grief and mourning, especially making use of special days (e.g., anniversaries).

Secondly, social workers as policy advocators

Social workers also advocated on behalf of survivors to help shape reconstruction policies and practices, helping government officials understand the importance of supporting the established social relationships and networks of families and relatives, neighbourhoods, and communities in the resettlement and reconstruction processes.

Thirdly, social workers as community organisers

Most social workers in these centres used a community organising approach to work directly with survivors and to empower them to actively participate in rebuilding their lives. Grounded in the community, the social workers and residents got to know each other better, so that the social workers could identify local expertise, familiarise themselves with the communities, and promote community cohesion among the residents. Besides, training local volunteers and fostering local leadership was one particular strategy they used most during their community mobilisation, with the purpose of deepening the capacity of communities and ensuring achievement sustainability.

Fourthly, social workers link economic development with social recovery

Economic development was the issue that most residents in the earthquake-affected areas cared most about. Tackling this issue helped social workers make real connections with the local people, in turn improving victims' emotional wellbeing by promoting economic recovery. The social service centres attempted to link economic development and social recovery through some strategies, such as running the handicraft producing project, providing microfinance to alleviate poverty and helping the village residents establish cooperation groups. But all of these endeavours were new for social workers and some of them clearly articulated that lacking knowledge of economic development issues brought huge challenges to their daily community work.

Conclusion and Lessons Learned

The Wenchuan earthquake was a defining moment and a transformational experience for the social work profession in mainland China. As a new profession in China, social workers started to establish their strong professional identities, launched massive social work practices in the earthquake-affected areas, articulated their roles in different stages of post-disaster interventions, and accumulated their best indigenous practices and experiences. People, communities, and the government, at local and central levels, recognised their function of disaster intervention and acknowledged their efforts as a new helping profession. Through the succinct summarisation of the involvement of Chinese social workers after the Wenchuan earthquake in this contribution we believe that the following lessons can be learned from their experiences:

1. Social workers in Mainland China lacked the knowledge and experience to respond to a natural disaster like the Wenchuan earthquake. Instead, social workers from Hong Kong and Taiwan were more prepared because of their prior experience in other disasters. The Wenchuan earthquake accelerated the development of post-disaster social work interventions in mainland China. Immediately after the earthquake, Chinese social workers were very eager to learn how to help disaster survivors. Such ad hoc reactive enthusiasm should be channelled to systematic capacity building to prepare Chinese social workers for future disasters. In sum, both Shanghai and Guangzhou social workers can use their experiences from the Wenchuan Earthquake to effectively deal with local disasters.
2. There is a strong need to develop manuals and practice guidelines for post-disaster interventions in Chinese, and to combine disaster mitigation and intervention as part of mainstream social work curricula in China. Chinese social workers found out that there were very few manuals and practice guidelines that they could use to guide their actions. Although many disaster-related documents, such as Psychological First Aid and the FEMA field manual have been translated into Chinese, they may not be applicable in China. Systematic research needs to be conducted to develop indigenous approaches toward disaster interventions. It is also important that disaster mitigation and intervention be part of the mainstream social work curriculum, as colleges in mainland China are developing social work

educational programs. In 2011 the Ministry of Civil Affairs published a book on disaster social work studies, Sun Yat Sun University published a book on disaster social work practice (Zhang et al., 2011), and Wang and colleagues (2012), from Beijing Normal University, edited a social work casebook of the Wenchuan Earthquake. Taken together, these efforts contribute to disaster social work curriculum design within the current social work education system.

3. The system perspective is unique to the social work profession, as we place an emphasis on understanding human needs and behaviours in the context of their social and physical environments. We found that this perspective is extremely useful in understanding how disaster affects the lives and well-being of survivors. Social workers can enhance the wellbeing of disaster survivors by managing the development of new support systems through advocacy and coordination roles. Social workers are also trained in a system perspective to understand the collaboration between different professions, as well as different institutions (e.g., different levels of government, NGOs, schools, and health care systems). Social workers are well equipped to provide case management and/or coordination among these different parties.
4. One major difference between social work and other helping professions is that most social work interventions are grounded in the community. Social workers in the disaster areas lived with the survivors in the temporary housing areas. They ate with the survivors and used the same public facilities. When compared to other helping professionals, social workers were easily the group most accepted by earthquake survivors. When other helping professionals like doctors and psychologists left, the social workers remained as companions of the survivors. Taken as a whole, the social work profession is not represented by heroic performances, rather, it is the acts of persistence and companionship that qualify social workers as helping professionals and, arguably more important, wins the hearts of the people.

References

Banerjee, M.M., and Gillespie, D.F. (1994) Linking disaster preparedness and organizational response effectiveness. *Journal of Community Practice, 1*(3), 129–142.

Chow, A.Y., Chan, C.L.W, and Ho, S.M.Y. (2007) How social sharing of bereavement experience by Chinese bereaved persons in Hong Kong. *Death Studies, 31*, 601–618.

Dodds, S., and Nuehring, E. (1996). A primer for social work research on disaster. *Journal of Social Service Research, 1*, 27–56.

Feng, Y. (2008) *Social work during the Taiwan 921 earthquake*. Retrieved 6 May, 2009, from http://www.kahabu.url.tw/workstation/03_reconstrac.

Ku, H.B., Ip, D., and Xiong, Y. (2009) Social work in disaster intervention: Accounts from the grounds of Sichuan. *China Journal of Social Work, 2*(3), 145–149.

Liu, Z. (2009) Post-disaster reconstruction and social work development in the Chinese mainland. *China Journal of Social Work, 2*(3), 221–223.

Lum, T.Y., Wang, X., and Danso, K. (2010) 2008 Wenchuan earthquake in China: Reflections on the role of social workers in disaster intervention, in *Disaster concepts and issues: A guide for social work education and practice* (205–220), edited by D.F. Gillespie and K. Danso. Alexandria, VA: CSWE Press.

Miller, J. (2012) *Psychosocial capacity building in response to disasters*. New York, NY: Columbia University Press.

Pyles, L. (2007) Community organizing for post-disaster social development. *International Social Work, 50*(3), 321–333.

Sim, T. (2009) Crossing the river stone by stone: Developing an expanded school mental health network in post-quake Sichuan. *China Journal of Social Work, 2*(3),165–177.

United Nations Economic and Social Council, (2008) "*Implementation of the Hyogo Framework for Action in Asia and the Pacific: Case study: The national disaster management system of China and its response to the Wenchuan earthquake*", available at: http://www.unescap.org/idd/events/cdrr-2009/CDR_2E.pdf (accessed on April 22, 2011).

Wang, X. (2008) Functions and roles of social workers during community reconstruction in the earthquake affected area, in *Report: In response to Wenchuan earthquake: policy recommendations* (in Chinese), 34–68, Ministry of Civil Affairs, China.

Wang, X. et al. (Eds.) (2012) *Starting from ground zero: Social work casebook of the Wenchuan Earthquake.* Beijing, China: China Society Publishing House.

Zakours, M.J. (1996) Disaster research in social work, in *Research on social work and disasters* (pp. 7–25), edited by C.L. Streeter and A.A. Murty. Binghamton, NY: Haworth Press.

Zhang, H. et al. (2011) *Disaster Social Work in China: Practice and Reflection.* Beijing, China: China Social Science Documentation Publishing House (in Chinese).

Recent Development on Disaster Management in Thailand

Decha Sungkawan

Disaster management has gained much public attention since the Indian Ocean Tsunami hit Southern Thailand at the end of 2004. A consultative meeting was held at UNESCO Bangkok in October 2005 where numbers of representatives of Thai public agencies and civil society organisations attended. Key recommendations from the meeting included the need for adequate access to localised disaster preparedness educational materials for local leaders and communities to use in disaster risk-training in disaster prone areas. Since 2005 Save the Children Sweden and partners have been working with 25 schools in tsunami-affected areas in Southern Thailand to build and enhance the skills and knowledge of children through disaster risk-reduction activities. The activity ranges from learning about different types of disasters to promoting disaster-preparedness in schools and communities. School principals and the teacher team have been effecting the inclusion of Child-Led Disaster Risk Reduction education into the schools' curriculum – this is designed for students in grades 4 to 6 and its objective is to empower students to apply local knowledge to solve problems in their daily lives. The curriculum is not officially part of the schools' curriculum but is an extra subject. The principals and teaching teams have suggested that it be incorporated into the schools' curriculum and be self-sustaining and be applied in a specific way to each school's needs. The curriculum would also need to be incorporated into subjects that are already being taught such as geography, social science, math and the Thai language, so that it would be easily accessible for students. Many students have been trained in the Child-Led Disaster Risk Reduction project and the curriculum will also be introduced to schools in other parts of Thailand with close monitoring and evaluation.

It was also concluded that a major seasonal disaster facing Thai citizens in specific areas is landslides. Landslides or mudslides are serious geological hazards that occur in 15 Northern provinces as well as in an additional 36 provinces around the country. The project was designed to produce an informational video to achieve the above-mentioned objectives for people living in landslide-prone areas in the northern region of Thailand. A video script was drafted aiming to tell the story of landslide hazards and preparedness in Chiang Dao District and Chiang Mai province in particular. The structure included an introduction, facts, and features, testimonials, call-to-action and conclusion. It was expected that in the closing stage, viewers would remember two keys points: "always stay alert" and "be aware of exposure to landslide risks".

Community-based and Gender Awareness Perspectives in Disaster Risk Management

Thailand adjusted its previous concept of disaster risk management from "emergency management" to "disaster risk management" when the country adopted the concept for disaster risk management agreed upon by the international communities in the meeting in Kobe, Hyogo prefecture in 2005 with the common principles of; the risk reduction of the potential disaster, disaster risk management awareness-building among community members; building awareness and understanding of the disaster and providing preventive and mitigation measures to the public; establishing preparation measures for disaster mitigation, and setting up operational guidelines for practice. The Community-Based Disaster Risk Management model provides the steps for practice accordingly; awareness-building among community members on the disaster risk they are facing, educating the community members on disaster risk prevention and mitigation through disaster risk assessment and analysis, forming community organisations to perform and function in disaster risk prevention and mitigation; setting in place community plans for risk disaster prevention and mitigation.

Under the values and ethical principles that govern social work practice, the Bureau of Women and Families Affairs, Ministry of Social Development and The Department of Disaster Prevention and Mitigation (Ministry of Interior) have issued the *Manual for Disaster Management: A Gender Awareness Perspective* as guidelines for social workers from both public and private sectors who are working in the areas of disaster management nationwide at different levels.

According to the manual, good practice for disaster management should be operated under the perspective of gender awareness in order to create gender equality among women and men who participate in disaster management. Women and men who are participating in disaster management as either service receivers or service providers should be treated properly and equally in all three phases of disaster management; pre-disaster, during disaster, and post-disaster. The gender awareness perspective in disaster management differentiates between sex, gender, and gender roles, thus they have different problems and needs. Male and female have different capacities and skills and both can contribute their endeavours and receive services according to their capacities and situations in different phases of disaster management. Gender awareness perspective in disaster management would also consider the rights and safety of women – particularly in the situation where numbers of men and women have to stay together in groups at a shelter – who should be protected from the violation of their rights and from threat while residing in the shelter.

Gender awareness perspective may be applied in disaster management at the community level in order to mitigate risk from the disaster. The perspective also promotes the capacities of male, female and other risk group members of the community. In the pre-disaster period, the gender awareness perspective focuses on knowledge and awareness building knowledge about disaster and risk management among community members. In the during disaster period, the community members are trained in how to respond to the crisis and emergency situation. In the post-disaster period, the community members are trained to participate in recovery and rehabilitation and sustainable development.

Community-based disaster risk management and the gender awareness perspective share certain values and ethical principles with the social work profession, particularly in regards to community work and case work methods. These two frameworks enable the social workers to understand disaster hazards and vulnerabilities as well as the capacities of human beings and to assess the effectiveness of social work practice. Social workers, as well as the other professionals who are engaged in risk disaster prevention and mitigation organisations at both the national and local levels, are generally accepted and integrate the community-based risk disaster management and gender awareness perspective frameworks into the development of their models of practices for disaster prevention and mitigation.

Social Workers' Role in Disaster Risk Management

As with some other countries in the Asia-Pacific, one of the areas most prone to natural disasters, Thailand has long and extensive experience in natural disasters management.

In response to the natural hazards, a small disaster mitigation unit was initiated within the Thailand Department of Public Welfare in the 1950s so social workers have played a role in post-disaster service delivery for several decades. The focus on immediate and short-term mobilisation of resources to protect life and property in the areas hit by the natural hazards has, to some extent, limited the development of professional capacity among social workers. Based on the Disaster Prevention and Mitigation Act 2007, disaster risk management in Thailand has now been developed and escalated in response to higher level of public concerns over more frequent severe disasters. Disaster risk management has been expanded to include planning and preparation phases and is now organised under a high level of governmental structure with local networks throughout the country. An interdisciplinary disaster risk management team has been appointed to address the various types of hazards in different phases. As a part of the disaster risk management team Thai social workers had been assigned to function mainly in post-disaster service delivery for half a century and this limited and passive role in disaster services needs to be redefined and their roles extended to fit current needs.

Social workers in disaster management have a primary function in responding to the impact and needs of disaster victims. The social work perspective in disaster services and disaster victim rehabilitation is specified in the Finance Ministerial Regulation on Payment of the Government Reserve Budget for Emergency Assistances to Disaster Victims 2003. Under this Regulation social work in disaster management has been given a specific definition;

> Social work perspective in disaster services and disaster victim rehabilitation means a short-term assistance to the disaster victims in order to strengthen them to be able to live on their own after the disaster. The disaster services provided by social workers focus on particular target groups such as children, elderly, disabled of which head of the household had passed away, disabled, or injured caused by the disaster. The disaster services may include removing of family from the disaster ground and providing a short-term vocational training for the family of disaster victims. The services may also include counsel services as well as referral services in order to meet the needs necessary for fast gaining of independent living.

Social workers who provide disaster relief services for disaster victims are classified into three major levels according to their experience and skills in this area of practice; case worker at the frontline level, social worker at the supervisory level, and the social worker at the administrative level.

At the administrative level social workers are normally assigned to provide service as the case worker with basic responsibility for note-taking, interviewing, and case investigations for service provision. The case worker is also assigned to deliver service to the victim's family members, arrange home visits, and provide counselling services for the victims. Social workers at this level are also expected to take responsibility for case recording and collecting all service data statistics. Social workers at the supervisory level provide care services to clients just as the frontline case workers do but they are also responsible for planning, coordinating, case assignment, supervising, case reviewing, and evaluation of the practices of the frontline social workers concerning post-disaster service and management. Social workers at the administrative level are responsible for reviewing the disaster victim service plans as well as conducting research and evaluation of the service system provided by the social workers – both the frontline case workers and the supervisory social workers. The administrative social workers may suggest changes and improvements in service practice and systems through training programs. The administrative social workers participate in programming and planning on disaster risk management with the interdisciplinary team. The administrative social workers also participate in policy processes on matters concerning disaster risk prevention and mitigation, such as personnel and budget.

Disaster Management in the Relief Shelter for Flood Victims at Thammasat University in 2011: Social Work Perspective

I will now turn to the more explicit function of social work practice in disaster management, referring to the flood relief shelter at Thammasat University, Rangsit Campus in Pathumthani Province north of Bangkok which is a part of a large-scale flood relief following the disaster that hit the central part of Thailand from July–October 2011 – The shelter opened for service on the first weekend of October 2011. The campus is located on the lower part of the central plain where the whole area has been hard hit by a number of typhoons and the torrential rains that caused large-scale flooding for months. The campus was protected from the flood water by the mud walls that were built around it and was one of the few safe places in the area – suitable for shelter for the flood victims whose houses had been under water for weeks (for months in some cases) since the beginning of flood in July 2011. The capacity of the flood relief shelter on the Thammasat campus was 4,000 and it had accommodated over 3,800 flood victims who were native Thais and foreign workers from the central part of the country, particularly residents from Ayudhaya and Pathumthani provinces as well as numbers of foreign workers from nearby industrial parks.

The student volunteers in the shelter were recruited from the university and included social work students. They provide various kinds of services for the flood victims which include victims' registration and welcoming, providing material services, communication, donations, and group activities for specific target groups, child protection, food and kitchen services, and cleaning services. The student volunteers are trained by specialist, experienced trainers before being assigned to specific services.

The student volunteers are trained to provide services as part of their learning experience through the service learning program that has been in place at Thammasat University since the 1970s when the graduate volunteer program was established.

In order to integrate the values and ethical principles that govern social work practice into the context of disaster management, Thammasat social work faculty members and social workers from both government and non-government social service agencies, under the umbrella of the Thailand Social Workers Association, have formed a network and set up the plan to provide flood victims with service and protection while they are in the shelters. One hundred and fifty social work student volunteers have been trained by experienced social workers from both the faculty and service agencies to provide four specific services in the shelter;

First, the demographic database of the flood victims was drawn from the shelter service registrations. The flood victim population was mixed in terms of residents, occupation, socioeconomic status, education, skill, and culture. The characteristics of the demographic variety reflect the heterogeneity of the communities in the lower central plains of Thailand. The flooding areas were considered as diverse communities that cover both old farming communities and new industrial communities where numbers of residents have recently migrated into the areas from both within and outside the country. The geographic database thus was needed in order to design the service programs for the flood victims while they are residing in the shelter as well as for post-disaster recovery.

Second, the protection service targeting specific risk groups including children, women, the elderly, disabled and infirm persons, and migrant workers: The protection services could be translated into various kinds of activities in the shelter. For child protection services the social worker could simply provide what is needed through group activities or events for children and teenagers according to their needs.

Third, the psycho-social counselling or case work interview service for those victims who have immediate needs for such services. Most of the flood victims were in the stressful situation of being flooded for more than a month before moving into the shelter. Many of the victims managed their stress through alcohol consumption and smoking. The stress situation for this group of shelter residents was more obvious when alcohol consumption and tobacco smoking were not allowed in the shelter. A mental health assessment form was thus in need for referral and service design.

Fourth, a needs and demands survey for post-disaster response and recovery. This survey was later conducted by the social work students who took their field work practical in the flood-affected areas, and at the agencies that provide post-disaster relief services.

These four specific services are provided for both the native Thai flood victims and the migrant workers who are accommodated in a separate shelter in the compound at Thammasat University, Rangsit Campus.

In order to provide protection services for specific risk groups, the social work student volunteers are trained to give help to children and families through various types of group activities depending on the interests and demands of the groups as well as the available resources for such activities. The child protection activities are generally welcome by the shelter management and the families of the children who regularly participate in the activities.

The case work interview service and the survey on needs for post-disaster response and recovery faced challenges among the actors in shelter management at the beginning as some key actors consider the services as a disturbance and violation of the rights and privacy of the flood victims. These services have therefore been delayed in order to allow people to settle in to the shelter.

The social work network for the temporary shelters for flood victims at Thammasat University also expands coordinating functions in supporting social work practices within and with networks and agencies outside the shelter. These coordinating functions include; coordinating bodies with shelter management structure, with donors, with volunteers and networks, and with other shelters.

In the post-disaster period it has been decided by the social work faculty members to place social work students in the communities around the campus as well as in other communities in the lower central part of Thailand as part of their field work practicals. The students are expected to assess the socio-economic impact of flood disaster and find out ways to reduce vulnerability in the flood-prone communities.

In trying to integrate the values and ethical principles that govern social work practice into the context of disaster management at Thammasat University's relief shelter for flood victims, the social work network; professional social workers, social work educators, and the social work volunteer students face a number of challenging issues.

First, the issue of misconception on the contributions and roles of social work in disaster risk management. Social workers in the Department of Disaster Prevention and Mitigation are assigned for functions limited only to short-term relief by providing and distributing necessities to the disaster victims. This function was established when the unit of disaster prevention and mitigation was founded and it is a typical role of social workers in most disaster prevention and mitigation units around the country. The advocacy role of social workers in the context of disaster management is not yet well established social worker is still limited. The social work network has difficulties in reasoning their other roles and functions – particularly advocacy for the risk groups such as children, women, and elderly victims – to other disaster management partners.

Second, the administrators of disaster management are preoccupied with their operating and rescuing functions and tend to see the other professionals and volunteers, including social workers and social work student volunteers who do not directly contribute to operating and rescuing functions, as less effective partners in disaster management. The professional social workers and social work student volunteers have difficulty presenting their argument particularly in emergency situations as the disaster management administrators assign all the volunteers, including social work student volunteers, to the functions such as rescue out-reach teams to supply food and drink to those victims who are isolated by the flood, and operating functions such as procuring donations, cleaning, washing, and kitchen jobs. The social workers and social work student volunteers thus often lose the chance of providing their professional services to the victims who need them.

Third, professional social workers and social work student volunteers in the flood victims' shelter face the issue of ethical dilemmas while they are trying to secure information by interviewing and surveying in order to provide their services and protective services to vulnerable risk groups and individuals flood victims such as children, women, infirm persons, and migrant workers – they have to secure information in such a manner that they are not intruding on privacy or abusing information they are taking from the flood victims. This ethical dilemma means that basic and protective services are provided only to those who need them by chance as the social workers and student volunteers are not allowed to secure information from all of the flood victims.

Fourth, social workers and social work student volunteers have discussed their mission and roles for the flood victims at the shelter and these include tasks beyond post-disaster service delivery. After running the shelter service for a while their daily debriefing covered the need to expand the post-disaster service to those victims residing in the shelter. *The survey form and interview guidelines have been set up for the task of post-disaster service delivery which may involve different service agencies both at policy and practice levels to response according to the needs of the victims*. Surveys and interviews for post-disaster service delivery, however, are not welcomed by shelter management authorities who focus their concerns on shelter operating and management and give low priority to tasks beyond shelter relief.

Fifth, in running the relief shelter for the flood victims the shelter authorities have to look for donors. There are many donors who offer their sponsorship to the shelter but social workers and social work student volunteers sometimes run into conflict with donors or sponsors as some activities that are created to provide services to the flood victims somehow overlook the values and ethical principles that social work applies to protect and advocate for the victims. There are some service program activities in the shelter designed by and controlled by the donors such as sponsorships from visual media industries that focus on media program production which may impinge upon the rights and privacy of the victims.

Sixth, the issue of environmental imbalances between the flood-prone area and other areas not being caused by natural factors but, somehow, being concerned with the politics of disaster management, where conflicts between group interests occur due to access to services and protection during and from disasters. These conflict situations could also produce environmental imbalances when some areas obtain more services and greater protection than others. The environmental imbalances, whether caused by natural factors or political factors, have implications for the practice of social work and social development in disaster management.

Although social work education and practice have been developed in the country for more than half a century and the Social Work Profession Act 2013 – which requires all social workers to be properly trained in each area of practice – the social worker's role, particularly in the area of disaster management, is still far from gaining full recognition in the region or Thailand as a whole.

Conclusion

Disaster risk management as a concept is rather new in Thailand but for more than half a century social workers have participated in disaster management and they have been assigned mostly to post-disaster relief with fewer opportunities for their professional development. As some other parts in Asia-Pacific, Thailand is now facing more frequent natural disasters and large numbers of the population are vulnerable to disaster risk and prone to be disaster victims. Children, the elderly, and heads of households are among those who passed away, were disabled, or injured by disaster. The comprehensive disaster risk management that covers the essential elements of preparedness, mitigation, emergency response, and recovery need to be applied in the situation. With knowledge and skills in human behaviour, particularly in the context of changes in environment, the social work profession could offer a range of services for disaster victims who have immediate as well as long-term needs.

References

Asian Disaster Preparedness Center (ADPC), (April 2008). *Mainstreaming Disaster Risk Reduction in the Education Sector in Lao PDR: Integrating Disaster Risk Reduction in the School Curriculum, Impact of Disasters on the Education Sector, School Construction: Current Practices and Improvements Needed.* Bangkok ADPC.

Faculty of Social Administration, Thammasat University, Summer2011 *Manual for Field Work Practicum* 1 (Thai Language).

Faculty of Social Administration, Thammasat University, Occasional Paper (January 2012). *Social Protection in Natural Disaster Management* (Thai Language).

Faculty of Social Administration, Thammasat University, Occasional Paper (November 2011). *Management in Temporary Flood Victims' Relief Shelter at Thammasat University,* Rangsit Campus (Thai Language).
Harvey, P., K. Haver, A. Harmer and A. Stoddard, (February 2010) *Humanitarian Coordination in the Asia Pacific Region*: Study OCHA Donor Support Group Field Mission, United Nations Office for Coordination of Humanitarian Affairs.
Royal Thai Government Gazette, (January 2013) *The Social Work Profession Act 2013 Ratchakitchanubeksa* (Thai Language) www.ratchakitcha.soc.go.th.
Tokiwa International Victimology Institute, (October 2005). Disaster Victimization: Understanding and Responding to the Impact and Needs of Disaster Victims. Prodeeding of the 3rd Symposium of the Tokiwa International Victimology Institute, October 13–14.
United Nations Report. (26 October 2010) *Asia-Pacific Most Prone to Natural Disasters* www.un.org/ … /news/story,asp? News ID … Cr=disaster.
United Nations (2011) *Global Assessment Report on Disaster Risk Management* Revealing Risk, Redefining Development. International Strategy for Disaster Reduction.
World Bank and Australia Aid, (2012) *Building Urban Resilience: Managing the Risks of Disasters in East Asia and the Pacific: Principles, Tools and Practice Disaster* Risk Management Team, East Asia and Pacific Infrastructure Unit.
Yodmani, S. (2001) *Disaster Risk Management and Vulnerability Reduction: Protecting the Poor.* Paper Presented at the Asia and Pacific Forum on Poverty Organized by the Asian Development Bank- the Social Protection Workshop 6: Protecting Communities-Social Funds and Disaster Management. United Publication.

Humanitarian Aid in Times of Disasters: Lessons from Responses to the 2004 Tsunami in Sri-Lanka

Tom Vickers and Lena Dominelli

Introduction

The Indian Ocean Tsunami of 2004 devastated lives in 12 countries, causing 300,000 deaths overall. This contribution focuses on Sri Lanka, where immediate aid was provided primarily by local people. However, the tsunami generated a huge groundswell of goodwill across the world and people, money and other resources poured in to assist those in need (Harris 2006). International interventions to deliver aid and provide humanitarian assistance have been criticised as often involving exploitative and oppressive practices (Hancock 1991; Hoogvelt 2007). To examine this contentious issue, the ESRC funded a study between 2009 and 2012 to investigate two international initiatives that claimed to respond to the tsunami in Southern Sri Lanka through empowering practice. Both are ongoing and have evolved to assume new directions. To maintain anonymity, we refer to them as the Institutional Model (IM) and the Professional Practice Model (PPM).

Methodology

Our study explores the contested nature of international interventions using Berger and Luckman's (1967) interpretative ethnographic approach that views reality as socially constructed, giving the researcher the aim of understanding the multiple social constructions of meaning and knowledge. Ethnographic methods used include participant observation, conversations and interviews of villagers of all ages, village elders, officials, staff and students providing aid that sought to empower local people. These were supplemented with a mapping exercise of donor NGOs; an on-line NGO survey; and focus groups. The research team included a principal investigator, co-investigator, research associate and secretary based in Durham, and three research associates based in Sri Lanka. The research yielded 386 transcripts (interviews and focus groups), 35 sets of field notes, and 45 questionnaires. A grounded approach was taken to analyse this data, and develop a thematic framework through an iterative process, so that themes emerged from the data instead of being pre-imposed (Mathbor 2008: 37). For a fuller discussion of our methodology, see Vickers and Dominelli (in press).

The findings from this project were wide-ranging, and only a few are summarised here (*see Dominelli (this volume*), Dominelli (2012) for further findings on gender, and Vickers and Dominelli (in press) for impact on students). In this paper we offer an overview of the international partnerships involved in the IM and PPM models, and propose key features for empowering partnerships that are significant for social workers intervening in post-disaster situations. Both models responded to the immediate emergency, but went further by developing links with local organisations and institutions to create strong, cross-border partnerships that have proven sustainable that are fully operative today.

Two Models for International Partnership

The IM began with senior management in a British university responding to the tsunami's widespread destruction of school and pre-school buildings with replacements. Pre-schools and multi-use community centres have been built with funds raised by UK university students, many of whom subsequently spent summers teaching English and sports and engaging in cultural exchanges in Sri Lankan villages and universities. The IM shifted and diversified its activities over time, to encompass academic partnerships and capacity building. Benefits of the model have been many and varied. Those most valued by villagers included: physical resources, particularly buildings for community use; opportunities for contact with people from other countries; and improving their English. A Sri

Lankan housewife in her forties, who lost five family members in the tsunami, told us how resources that came initially through the IM were adopted by the community and made their own:

> Children have a preschool. The school-going children have a library to read books and gain more knowledge … the villagers now speak a few words of English. They remember how friendly the students were. They remember the students and [senior project staff] for giving the building to the village. They are also thankful for giving the children scholarships … The teachers are now very attentive about the children. They also take care of the maintenance of the building.

The PPM was launched by a civil society organisation that supports, networks and builds capacity in social work education at higher education institutions internationally. Social work training in Sri Lanka has been provided by a state institute since 1952, training limited numbers of students. Following the 2004 tsunami, social work educators and students from universities including the UK, Slovenia, Canada and Hong Kong engaged with Sri Lankan universities, government agencies and NGOs, to support interventions in villages directly affected by the tsunami, providing immediate aid. This work later addressed capacity building in social work education and practice which lies at the heart of the PPM. The Sri Lankan villagers valued the opportunity to speak to a stranger about their experiences and emphasised the activities students conducted to help children cope with fear of the sea. One Sri Lankan fisherman from a village where the tsunami killed 70 people claimed:

> the support given by the students was very welcome at a time that we were sad. They came, they stayed with us, worked with us. They helped us … even gave the people financial support, material support and … money and medicines. Most of all, the fact that they were with us.

Social work educators from overseas and from Sri Lankan universities doing this work in the field realised the importance of building capacity in social work education and community development to reduce risk from future disasters and deal with current social problems, many with roots in the tsunami and began to concentrate on these. The PPM emerged through this process rather than from a pre-planned model. Staff and students from a Slovenian university and Sri Lankan university have spent considerable amounts of time at each other's institutions, attending and giving lectures and undertaking practice placements. Social work educators from Slovenia, as well as others from the UK and Canada have contributed to the development of social work curricula at Sri Lankan universities and institute for social work.

Key Features of Empowering Interventions

The ESRC research found that both models: formed intensive and long-lasting partnerships built on trust; listened to local actors; and engaged local communities in empowering processes of giving and receiving aid. Although these elements were not always achieved, these models contributed to empowerment when they:

- Had clear aims and objectives, widely understood by people working to implement them.
- Responded to psychosocial and material needs.
- Supported local people in becoming self-sufficient.
- Facilitated exchanges of practices that were suited to local ways of doing things, using available resources.
- Included opportunities for sustained contact between diverse peoples, consciously aiming to improve cultural understanding.
- Involved local partners with skills, networks and other resources that could be supplemented to fill gaps.
- Were supported by individual relationships between members of institutions at all levels.
- Acknowledged power inequalities and potential for conflicts, and built structures that took these into account.
- Were made within a long-term perspective involving sustainability and accountability.

Conclusion

The ESRC research indicated that there are models of intervention that demonstrate that important benefits can arise from genuine and careful engagement with local people that empowers them in determining locality-specific and culturally-relevant responses to disasters. The IM and PPM are two such models.

References

Berger, P.L., and Luckmann, T. (1967) *The social construction of reality: a treatise in the sociology of knowledge*, London: Penguin.

Dominelli, L. (2012) 'Gendering climate change: implications for debates, policies and practices' in *Research, action and policy: addressing the gendered impacts of climate change*, edited by M. Alston and K. Whittenbury. New York, NY: Springer.

Hancock, G. (1991) *Lords of Poverty: The Power, Prestige, and Corruption of the International Aid Business*. New York: Atlantic Monthly Press.

Hoogvelt, A. (2007) 'Globalisation and Imperialism: Wars and Humanitarian Intervention', in *Revitalising Communities in a Globalising World*, edited by L. Dominelli. Aldershot: Ashgate.

Harris, S. (2006) *Disaster Response, Peace and Conflict in Post-Tsunami Sri Lanka, Part 1: The Congestion of Humanitarian Space*, Bradford: University of Bradford Department of Peace Studies.

Mathbor, G.M. (2008) *Effective community participation in coastal development.* Chicago, IL: Lyceum.

Vickers, T. and Dominelli, L. (in press) 'Students' involvement in international humanitarian aid: learning from student responses to the 2004 tsunami in Sri Lanka', *British Journal of Social Work.*

Invisible Women: Examining Women's Experiences of Aid in the 2004 Indian Ocean Tsunami in Sri Lanka

Lena Dominelli

Introduction

The Indian Ocean Tsunami of 2004 devastated lives and landscapes in 12 countries, causing 40,000 deaths in Sri Lanka alone. Gestures of goodwill generated substantial humanitarian aid, but raised critical questions its delivery and receipt. Contentious critiques alleged that donors benefited more than recipients (Hancock 1991, Hoogvelt, 2007). This concern was explored in research entitled *Internationalising Institutional and Professional Practices* (IIPP) (1), which sought to answer two questions through ethnographic research that highlighted the voices of Sri Lankan aid recipients in two models that purported to empower aid recipients – the institutional model (IM) and Professional Practice Model (PPM). These questions were:

- Are some models of humanitarian aid better than others?
- How do class, gender and ethnic relations impact on aid processes?

The IM was initiated under the auspices of a British university; the PPM arose through a voluntary professional organisation promoting social work education internationally. Both models began by providing assistance in the immediate aftermath of the tsunami and shifted into capacity building in local communities during the reconstruction phase in response to local residents' wishes. Both models are ongoing. The names and numbers of those involved have changed over time, although a core group remains active in each model. The data generated by the IIPP is extensive and archived for others to mine. In this volume, both this contribution and that by Vickers and Dominelli *(pages)* draw upon it. Although each contribution focuses on a different theme, there will be some overlaps around context and methodology.

Conducted during 2009 to 2012, the research sought answers to the above questions by focusing on Sri Lankans' reflections upon past experiences. In this contribution, I concentrate on expressions of gender relations during the tsunami, within Sri Lankan family relationships; between aid donors and recipients; and women aid workers. Most research ignores the specificities of women's experiences (Enarson and Morrow, 1998; Enarson and Chakrabarti 2009). This contribution contributes to addressing this knowledge gap.

Methodology

The IIPP endeavoured to evaluate two aid models in which donors claimed to empower local people and covered 12 villages in which the institutional model (IM) and professional practice model (PPM) were practised (Vickers and Dominelli, forthcoming). The research team included a principal investigator, co-investigator, research associate and secretary based in Durham; and three research associates based in Sri Lanka, one of whom was a woman. As the research crossed borders, cultural and linguistic sensitivities were addressed by the inclusion of local researchers as full team members responsible for conducting on-site research.

The research drew heavily on ethnographic methods to access rich data that would shed light on people's experiences of humanitarian aid as they perceived them. The methods utilised were:

- ethnographic methods including participant observation
- conversations and interviews of villagers of all ages, village elders, officials, and staff and students providing aid claiming to empower local people through the IM and PPM models
- a mapping exercise of donor NGOs
- an on-line NGO survey; and
- Four focus groups

The research yielded 386 transcripts (interviews and focus groups), 35 sets of field notes, and 45 questionnaires. NVIVO was used to analyse this data, and a thematic analysis of gender relations was employed for this paper.

Despite the substantial data generated, this research did not seek to establish externally validated truths about Sri Lankans' experiences of aid. Rather, it explored social constructions of their realities and the patterns depicting the range of verities presented by respondents (Berger and Luckman 1967). Although the data has internal validity, the findings have to be interpreted cautiously and not over-generalised.

Findings

The project's overall findings revealed that people generally appreciated the aid offered through the two models examined. The gender specific thematic analysis highlighted the invisibility of women as gendered subjects, replicating findings highlighted earlier by Enarson and colleagues. This study indicated that gender relations were socially constructed according to cultural expectations about acceptable behaviours in what women and men in Sri Lanka can and cannot do. Gender, unremarked upon, created a 'gender silence' and invisibility through which gendered social constructions affected women's opportunities and choices whether it was while obtaining aid immediately after the tsunami, during recovery, or reconstruction phases. Being restricted in what they could expect and do complicated women's lives.

Gendered expectations acted as opportunities and/or barriers, but were embedded in culturally relevant gendered assumptions held by participants. Some NGOs specifically targeted women because they realised they needed gender specific help to overcome disadvantages produced by the tsunami whereby women had lost husbands or fathers responsible for their wellbeing, but without challenging traditional gender relations. For example, Herath (2008) describes how Kantha Shaktha, a local NGO working with Sri Lankan women engaged them in income generating activities. As their confidence increased, the women began to exercise agency and organise activities with other women such as visiting the Temple of the Tooth in Kandy. They would never have done this before, but were careful not to upset their husbands. An interesting finding in the IIPP study was that gender relations were configured around everyday life practices that linked women's daily routines of caring for children, relatives and husbands, doing housework and cooking. Thus, everyday life practices created spaces in which women could exercise agency while simultaneously remaining within culturally bounded gender relations.

Gender relations within Sri Lankan family relationships

Traditional gender relations in Sri Lanka were patriarchal. This privileged men, but identified specific spheres of activity in which each gender flourished. Maintaining community harmony was an obligation for all residents – men, women and children. The economic division of labour was central to the social construction of gender relations. Men were typically configured as economic providers. Expectations about masculinity gendered men and framed community responses to their behaviour. In the Sri Lankan research sites, men's economic dominance came from their fishing activities. Catching fish was the basis of their livelihoods, and affirmed their masculinity. Both were dented when the tsunami destroyed their fishing boats, and life became particularly difficult if these were not repaired or replaced straightaway. Men's responses to this disaster varied. Some were positive; others undermined communal harmony. For men and their families, getting men back on the ocean waves to fulfil their responsibilities in providing for their families was an urgent priority. However, a few men had become concerned about risking their lives on the ocean and sought different means of generating an income. Other men responded in less socially acceptable ways by misusing alcohol and other substances. Such behaviour produced problems not only for these men and their families, but for the community as a whole. The following comment replicates what many felt about men who did not conform to traditional expectations of masculinity, especially amongst younger men:

> Our boys were very good, they helped others who came to clear the dirt, bury the bodies, and clean the houses. Some of the young people had lost their boats and fishing things and they were sad. Some bad things also happened. Some of the boys got to drink arrack and began to fight. Some boys became lazy.

This research revealed that differing expectations around gendered activities produced diverse outcomes for women and men, although both had been affected by the tsunami and had experienced extensive devastation and loss. This comment pithily encapsulated this view:

> If men's economic activities were hampered [by the tsunami], they are excused … .Women are not excused by the society. They are supposed to do whatever … deal with the problems.

Life after the tsunami was hard for all residents. However, women endured gendered forms of violence that were difficult to transcend, as this participant makes clear:

> Men coming home from fishing in the morning. They drink and sleep. That's all they do. Wife battering is very common.

Women specific resources to support them in these conditions were seldom available. Abiding by cultural norms and expectations was unavoidable. Some participants indicated that gendered cultural barriers were powerful and difficult to change. This also held if women had lost breadwinners but the family's fishing boat survived. The woman would was unable to assume the fishing role because strong taboos prevented her doing so, as this respondent clarifies:

> In fishing villages women are in a lower position because men are very dominant … women are not supposed to be on a fishing boat. It will get polluted … . so women were completely excluded.

Additionally, women were expected to pay the price personally to ensure that their families survived and received the care needed, as this respondent indicates:

> She didn't want to get married … take a job in a garment factory which she has to because her husband drinks too much … and she's got a small child … she wanted … to teach.

This woman could not live out her dream as long as her family circumstances prevented her from doing so. Her configuration as a wife and mother meant that meeting the obligations associated with these roles took precedence.

Women were also expected to act as chief mourners and this could impact strongly on everyday life activities as they conducted normal daily routines: 'Our children died … This happened to us, [women] are not supposed to laugh'.

Girls and young women were expected to live within the confines of gendered realities. Interestingly, this could present dilemmas for women aid workers who might experience conflicts between personal values and Sri Lankan ones. But, they felt unable to question or undermine traditional gendered ways of doing things, as described below:

> The girls don't play cricket on their own without female chaperones … what we're doing is cementing the gender divide that already exists.

Protecting family honour mattered enormously to people living in the research sites. Upholding gendered norms could lead women to choose death over life to prevent their behaviour from bringing shame onto their families, as this respondent indicates:

> Women and children were killed for social, cultural reasons … there was … this girl drowning and somebody gave her a hand to reach her … when she realized that she didn't have any clothes on … she let the hand go and she let herself drown.

The images conveyed through these narratives are moving. Much of their power derives from what is left unsaid. Gendered interactions within families were not explicitly defined as gendered by participants. Gender became visible through what they said about their expectations about what was permissible for men and women to do, and by how caring work was assumed to be just that. Caring work simply got done. Who did it was not a matter for comment. That it was women doing this work was evident to the researchers. Such taken-for-granted assumptions, evident in the following quotation, reveals that the caring work that women do remains invisible, even though its positive impact on well-being (of children in this case) was valued and appreciated:

> I feel that my daughter is better now, all her improvements are from the village pre-school … [T]his pre-school is improving, we have to protect it.

The people working in these pre-schools were exclusively women, but the respondent ignores its normality. Normative behaviour passes without comment.

Gender relations between aid givers and aid receivers

Gender relations are evident in every aspect of life, including in relationships between aid givers and aid receivers. Women distributing aid shared experiences encompassing the invisibility of their labour and 'gender silence' with women receiving aid. The researchers were surprised by the continued lack of visibility of women as gendered subjects amongst both donors and recipients. Although exposed by Enarson and Morrow (1998) 20 years earlier after surviving Hurricane Andrew in the USA in 1992, even the NGO community neglected women workers and women's specific needs as women receiving aid ... One woman aid worker commenting on women's invisibility in the NGO community said:

> I looked at ... gender and disaster management. I went to all the possible NGOs who had worked in these areas and I couldn't get any report ... with regards to that ... and ... it was women who were affected.

The gendered nature of social relations remains invisible in the distribution of aid to children as well as adults, as is suggested here:

> Some of the people who came [Red Cross/World Vision/FORUT/Christian Children's Fund] did some programs for the children ... They made them play games, made them draw, sing and dance ... Some organizations took the children to the beach.

The workers doing this work were mainly women. Other villagers commented positively on work that students had undertaken. But their gender was ignored, although most of them were women:

> It was good experience we got when we met the ... students ... .They sang songs; they worked with us to help build houses. We went to the beach and played on the shore. We had parties. They gave us nice flowers which made us happy and it was a new thing for us to give flowers.

Women aid workers' experiences suggest that aid donors take seriously women's issues and gendered relations if they are to avoid excluding women. Without their inclusion, communities cannot benefit from the many roles women can play in rebuilding community resilience, if given the opportunity.

Women aid workers did not set out to change gender relations in Sri Lanka. Some felt distinctly uncomfortable if they felt that others involved them in changing gendered customs and the existing gender balance, as the following woman student suggests.

> In the village we wore clothing down to our knees and t-shirts … some of the girls wore strap tops and really short shorts … it annoyed me because it wasn't respecting their culture … [A]t the university, the Vice-Chancellor said it was fine for us to wear shorts … He was almost encouraging us to … show [Sri Lankan] girls that they could … wear that, but I think they were quite shocked at us wearing that.

Gender specific relations affecting women aid workers

The position of women aid workers and discussions about their safety, sexualisation and abuse is missing from the aid literature. This research revealed this was a major issue for women aid workers. One described the sexual harassment that women aid workers experienced regularly on the buses as follows:

> If you're on a bus a lot of times … they would try and touch you and put their hands up our skirts and down your tops … they would stare at us … and made you feel really uncomfortable.

These women looked to traditional gender relations to solve this problem. One explained it thus:

> From a safety point of view, there … was a remarkable difference to be walking down the street in a group of girls and to be walking with one of the boys because of the attention that you attracted as a white girl in Sri Lanka.

Cross-cultural relationships involving women aid workers were problematised. One aid worker counselled against becoming involved in intimate relationships:

> [She] had a relationship with a Sri Lankan boy … and he hit her on the face … they ruined the reputation of our project … [and] of foreigners because Sri Lankans already have got the wrong idea about foreign girls.

Some aid workers felt strongly that interpersonal interactions should not undermine individual and institutional reputations.

Conclusions

Internationalising practices in aid situations are contradictory. Observing local cultures, traditions and languages, crucial to aid processes, can endanger women in various ways. As these are seldom discussed, they remain hidden under stereotypical assumptions and the social construction of traditional gender relations. This research revealed that stereotyped, unitary identities and traditional gender relations exclude women, during and after the tsunami. Realising equality and empowerment for women in such circumstances is difficult, and makes listening to women crucial to aid responses. Including women in capacity building endeavours from the word 'go' is essential in empowering gender relations and requires aid donors to promote:

- Effective and specific assessments of the different needs of women, men and children.
- Clear, transparent entitlement criteria, resource allocation and distribution systems and monitoring mechanisms.

Holding donors accountable for what they do and do not do was highlighted by villagers. From an equality perspective, ensuring that women's specific needs are addressed requires specific action from donors and recipients alike. They cannot be imposed by overseas actors but must develop locally.

Notes

Funding from the UK's Economics and Social Sciences Research Council and the entire research team contributed to this Project.

References

Berger, P.L., and Luckmann, T. (1967) The Social Construction of Reality: A treatise in the sociology of knowledge. London: Penguin.

Enarson, E. and Chakrabarti, P.G.D. (2009) Women, Gender and Disasters: Global Issues and Initiatives. London: Sage.

Enarson, E. and Morrow, B. (1998) The Gendered Terrain of Disaster: through Women's Eyes. Miami: Florida International University.

Hancock, G. (1991) Lords of Poverty: The Power, Prestige, and Corruption of the International Aid Business. New York: Atlantic Monthly Press.

Herath, S (2008) Women in the Sri Lankan Tsunami. Paper presented at the IASSW Congress in Durban, SA, 15–18 July.

Hoogvelt, A. (2007) 'Globalisation and Imperialism: Wars and Humanitarian Intervention', in Revitalising Communities in a Globalising World, edited by L. Dominelli. Aldershot: Ashgate.

Vickers, T. and Dominelli, L. (forthcoming) 'Students' Involvement in International Humanitarian Aid: Learning from student responses to the 2004 tsunami in Sri Lanka', submitted to the British Journal of Social Work (awaiting decision).

Social Workers as Crisis First Responders

Antoinette Farmer and Ronald Quincy

It is critical for social workers to assume leading roles as first responders during natural and man-made disasters. The National Association of Social Work (NASW) policy on disaster relief, clearly affirms this. The policy states that social workers are 'uniquely suited to interpret the disaster context, to advocate for effective services, and to provide leadership in essential collaboration among institutions and organizations' (p.83). Social workers are well equipped to assume leading roles as first responders because of their training in how to apply the ecological perspective to understand social problems; their knowledge of micro and macro interventions; and human development across the life span. All of the above are important for assisting persons who are experiencing trauma as a result of a natural or man-made disaster.

The purpose of this chapter is to discuss the knowledge and skills that social workers possess that equip them to be first responders, and the areas of disaster recovery where these skills can be used. Using a case example, it will be demonstrated how social workers from Rutgers, The State University of New Jersey were able to be first responders in the aftermath of the earthquake in Haiti.

Skills and Areas of Disaster Recovery where Skills can be used

Case management skills are essential when working with persons affected by disaster. These skills include advocacy, assessment, monitoring, and planning (Bell 2008). Advocacy can be used in assisting those persons get the services that they need to deal with the disaster. Monitoring can be used to determine if these individuals are receiving the services that have been identified as being needed. Social workers' knowledge and ability to navigate the various systems that persons affected by disaster may interact with are also important in disaster relief efforts.

Community revitalisation is an important aspect of disaster recovery (Pyles 2007). Social workers who are knowledgeable about community organisation can play a vital role in helping persons affected by disasters rebuild their communities. Given that disasters often present new challenges for vulnerable and disenfranchised populations (Pyles 2007), social workers can be involved in mobilising this population to address their needs through policy advocacy, developing grassroots organisations, and partnering with existing community organisations. It has been documented in the literature that inclusion of grassroots and informal organisations can enhance person's participation in disaster relief efforts (Harrell and Zakour 2000). Community organising efforts can focus on housing, community planning, and economic develop, as often times disasters affect the above-named domains.

Social workers who are knowledgeable about the effects of trauma can serves as members of the trauma response team. As part of a trauma response team, social workers may be involved in providing crisis intervention and grief and bereavement counselling. Moreover, social workers can help response team members recognise signs of distress, stress, and poor coping that individuals affected by disasters maybe displaying. Social workers can educate response team members about how stress can result in persons engaging in behaviours that they once did at a prior stage of development. Persons may engage in these behaviours as a way of coping with the trauma. Because social workers have been trained to assess for ones' strengths, they can work with persons who are affected by disasters to identify their strengths. These strengths can be used to help persons cope with what has occurred.

Social workers who have social work practice with diverse groups, both locally or internationally, or who are culturally sensitive maybe well suited to provide disaster relief in international settings. Disaster relief efforts by persons who are not cultural competent may cause more trauma to those already traumatised by the disaster (Thompson 2010).

Case example

On 12 January 2010 a 7.0 magnitude earthquake struck Haiti, killing 316,000 individuals, injuring 300,000, and causing \$7 billion dollars in damage (Farfel et al. 2011). It was estimated that 1 million people lost their homes; many of whom are still homeless today.

Social workers involved with the Haitians affected by the earthquake engaged in several aspects of case management. Prior to arriving to Haiti, social work researchers had begun planning how they were going to assess the needs of those affected by the earthquake. They developed a needs assessment instrument to collect data from individuals, families, and community leaders. In collaboration with Haitian indigenous individuals, the instrument was translated into Haitian Creole and tested for cultural competency. The needs assessment instrument was read to the persons being interviewed by an indigenous person working with the social workers from the United States. Geographic Information System (GIS) devices were used to collect the data. These data will eventually be used by the Haitian community partners to advocate for additional resources that will enable them to meet their identified needs. Therefore, the social workers enabled the Haitians to advocate at the macro level. Once the data are used by the Haitian community partners, the social workers will monitor if the individuals are receiving the needed services.

Conclusion

In conclusion, social workers are well equipped to be first responders in the time of disasters. Their ability to assume this role is based on their educational training. The social work curriculum, including the field experience, provides social workers with the opportunity to learn about micro and macro interventions, theories about how persons deal with trauma, and case management skills, which have been deemed to be important when working with persons affected by disasters (Bell, 2008). The skills and knowledge that social workers possess will allow them to be first responders in both local and international settings. With regards to being first responders in international settings, it is important that indigenous people are empowered to make decisions related to external relief, recovery, and community redevelopment and that one is culturally sensitive.

References

Bell, H. (2008) Case management with displaced survivors of Hurricane Katrina. *Journal of Social Service Research*, 34, 15–27.

Farfel, A. Assa, A., Amir, I., Bader, T., Bartal, C., Kreiss, Y. (2011) Haiti earthquake 2010: A field hospital pediatric perspective. *European Journal of Pediatrics*, 170, 519–525.

Harrell, E.B., and Zakour, M.J. (2000) Including informal organizations in disaster planning: Development of a range-of-type measure. *Tulane Studies in Social Welfare*, 21, 61–83.

National Association of Social Workers (2003) Disasters. *Social work speaks: National Association of Social Workers policy statements*, 2003–2006 (6th ed., 83). Washington, DC: NASW Press.

Pyles, L. (2007) Community organizing for post-disaster social development: Locating social work. *International Social Work*, 50, 321–333.

Thompson, G. (2010) Cultural competency: Haiti and at home. *British Journal of Community Nursing*, 15, 109.

Chapter 5
Indigenous Understanding of Environmental Change and its Social Consequences

Indigenous peoples are among the first to face the direct consequences of climate change due to their dependence upon, and close relationship and dialogue with, the environment and its resources. Indigenous peoples and the role they may play in combating climate change are rarely considered in public discourses on climate change.[1] Indigenous understanding of the world is holistic. All parts of the natural habitat are considered "sacred" and to be preserved for future generations. What can we learn from the indigenous experiences of surviving in a symbiosis with an ever changing natural environment?

Climate Change and Environmental Justice: Indigenous Perspectives from the United States

Hilary N. Weaver

My name is Hilary Weaver. I am Lakota, with family ties to Rosebud reservation in South Dakota, United States. I am also adopted into the Seneca Beaver clan and my Longhouse name is Ganoaset. My children and I reside within the Haudenosaunee territory in what is now the western part of New York State, United States. I have been asked to share my perspectives on climate change and Indigenous Peoples.

Human beings exist within a context that sustains and nurtures them. The resources found in the natural environment are what make life possible. Indeed, the environment has a shaping influence on cultures. Climate is an integral part of the natural environment. Whether a region is hot or cold, wet or dry; these factors are all significant to the plants, animals, and people found therein. All life adapts to the surrounding climate and environment. They exist within a symbiotic relationship. Traditional Indigenous Peoples, more than other people who have integrated into industrialised societies, maintain this close, symbiotic existence with the natural world.

While recent decades have seen an increased discussion of climate change and its impacts, there are still some people who continue to deny that climate change is real. Typically people who are likely to deny climate change are those ensconced in

1 Follow UNPFII United Nations Permanent Forum on Indigenous Issues (United Nations Permanent Forum on Indigenous Issues). [May 2013].

a position of privilege, far removed from the day to day rhythms of the environment. While some people remain oblivious to climate change, it will ultimately affect all of us. Climate change presents a threat to humanity at large and will have a devastating impact on indigenous people in particular (Williams 2012). This impact is already being felt by Indigenous Peoples in various regions of the world.

The initial impact of climate change is felt most keenly by those whose life and livelihood is mostly closely tied to the natural environment. In this respect, indigenous peoples who maintain a traditional lifestyle are among the most vulnerable to environmental changes. As plants and animals shift their locations or disappear along with their sustaining environments, the indigenous peoples who depend on them also face monumental changes.

The dramatic and rapid changes currently occurring in the natural environment are largely driven by human activities associated with industrialised societies. These changes require an equally dramatic and rapid response. While reversing climate change is highly unlikely, it is possible to recognise and mitigate many contributing factors. Responding to climate change is a matter of environmental justice (Trainor, Chapin, Huntington, Natcher and Kofinas 2007).

Our Changing Environment

Multiple examples of the world's changing climate can be found within the United States. Typically discussions on climate change focus on changes in the far north, and indeed flooding and erosion have become significant concerns as temperatures warm and the ice shelf melts in Alaska. The physical impact of climate change is most extreme in northern latitudes (Trainor et al. 2007). For example, between 2002 and 2007 coastal erosion more than doubled along a 40-mile stretch of the Beaufort Sea. As a result coastal villages were swamped. There have also been significant shifts in plant and animal populations as they try to adapt to thawing tundra (Weinhold 2010). This drastic example serves as a wake-up call to us all. The Arctic is sending a message that climate change is moving quicker than previously estimated (Goldtooth 2010).

The Arctic is not the only region in the United States facing climate change. Less often discussed are the effects of climate change in other regions. For example changing trends in spring thaw runoff in the South-western United States threaten agriculture and tourism. This, in turn, may lead to reliance on unsustainable natural resource extraction which further impacts the natural balance in the environment. In yet another region, decreased water levels and rising temperatures in the Great Lakes threaten natural species and encourage invasive species. Indeed, climate change not only impacts many regions but is a catalyst for additional changes in the natural environment.

The Impact of Climate Change on Indigenous Peoples

While the actions of industrialised societies are the primary causes of climate change, the toll that these changes take are much heavier elsewhere. The disruptions caused by climate change do not affect all populations equally (Trainor et al. 2007). Indigenous people trying to maintain a traditional lifestyle are particularly vulnerable. Climate change affects subsistence lifestyles including cultural and economic activities (Trainor et al. 2007); indeed, this affects our very existence as distinct indigenous peoples. For example, cultural ceremonies for Indigenous Peoples from traditional societies dependent on fishing traditions can no longer take place in the traditional manner when the fish needed for the ceremony have been displaced by invasive species. This lack of fish also has economic and nutritional impacts on traditional indigenous societies.

By definition, indigenous people are people of the land. Our cultures, social organisations, and spirituality are shaped by the territory we inhabit. Climate change raises key questions. What happens to us as Indigenous Peoples when the land or waters we depend on change? What is the impact when plants and animals that we depend on are no longer there and new species begin to dominate?

The impact of climate change cannot be overestimated. While many continue to ignore it, others recognise that it may ultimately mean an end to some traditional indigenous societies: 'The wanton destruction of traditional homelands of indigenous people and the extinction of their people by climate change does constitute cultural genocide and amounts to one of the most serious crimes in our human history' (Williams 2012: 649).

Climate change alters ecosystems that are the economic, cultural, and dietary cornerstones of indigenous existence. The existence and subsistence of traditional Indigenous Peoples is intimately tied to their natural environment. Depending on their surroundings, Indigenous Peoples have developed subsistence economies based on hunting, fishing, gathering and other aspects of their natural environment. After encounters with colonial societies, many indigenous groups preserved their rights to subsistence activities through legal agreements such as treaties.

These unique rights of indigenous peoples such as hunting and gathering are now affected by climate change: 'The way in which many Indigenous communities survive has much to do with their relationship to the land. When that land begins to transform and sustenance patterns lose their predictability, safety is at risk and Indigenous knowledge can be lost' (Lewis 2008: 12). Rapid transformations mean that generations-old knowledge about when and how to conduct subsistence activities no longer reflect current realities. For example thick ice that easily supports the weight of hunters may not exist when and where it has in the past. Under these changing circumstances, going out on the land presents new dangers.

Climate change has implications for food security and nutrition (Furgal and Sequin 2006). In Alaska, 80 per cent of indigenous peoples' diet comes from their natural surroundings (Weinhold 2010). Changes in those surroundings, including the scarcity of plants and animals that form the basis of a traditional diet, mean food insecurity, which in turn can lead to nutritional deficits.

Health is also affected by climate change. For example, climate change has led to more biting insects in the north with the potential to spread disease (Furgal and Sequin 2006). Changes in the environment have led to increasing water borne and vector borne illnesses (Hunter 2009). Water resources have been compromised thus affecting the supply of safe drinking water for both people and animals.

Around the world, Indigenous Peoples already suffer significant health disparities. Indigenous peoples' existing burden of ill-heath increases their susceptibility to the adverse impacts of climate change (Harries 2010). Additionally, there are unanswered questions about the impact of climate change on the mental health of Indigenous Peoples. Given that many indigenous people are already socially disadvantaged, they may experience increased risk factors for mental health concerns (Hunter 2009).

Indigenous Responses to Climate Change

While traditional Indigenous Peoples are interdependent with their natural environment they are far from passive observers of climate change. Many traditional teachings tell us that Indigenous Peoples shoulder a particular responsibility for the balance and wellbeing of the world (Goldtooth 2010). Traditional knowledge may well give us insight into solutions to address contemporary problems such as climate change.

Indigenous people from around the world have gathered to discuss climate change and have issued statements linking it to values of dominance over nature and the exploitation inherent in capitalism (–2011). Accordingly, some First Nations leaders continue to exert pressure on host governments to address climate change issues (Thompson 2010).

Indigenous Peoples now gather regularly to discuss climate change. On April 24, 2009 Indigenous people gathered in Anchorage Alaska for a Global Summit on Climate Change. They issued the following declaration:

> We express our solidarity as Indigenous Peoples living in areas that are the most vulnerable to the impacts and root causes of climate change. We reaffirm the unbreakable and sacred connection between land, air, water, oceans, forests, sea ice, plants, animals, and our human communities as the material and spiritual basis for our existence. We are deeply alarmed by the accelerating climate devastation brought about by unsustainable development. We are experiencing profound and disproportionate adverse impacts on our cultures, human and environmental health, human rights, well-being, traditional livelihoods, food systems and food sovereignty, local infrastructure, economic viability, and our very survival as Indigenous Peoples. Mother Earth is no longer in a period of

> climate change, but in climate crisis. We therefore insist on an immediate end to the destruction and desecration of the elements of life. (Goldtooth 2010: 18)

In spite of such impassioned pleas there has been an overall failure of governments around the world to adequately respond (Williams 2012). Ironically, while Indigenous Peoples are disproportionately affected by climate change, Indigenous governments may also fail to respond, often due to their already precarious economic standing and vulnerability (Weinhold 2010). For example, some indigenous governments that have adapted some industrialised ways have a narrow economic base focused on mineral or oil extraction and have become reluctant to give up this short term gain because it enables them to participate in the larger society's economy and thus purchase things like food and shelter for tribal members. Unfortunately this short term gain also contributes to long term problems with climate change.

Ideally what needs to happen are efforts where Indigenous and non-Indigenous Peoples can develop partnerships to address these issues. Canada has been open to including aboriginal leaders in dialogues on climate change (Furgal and Sequin 2006). Traditional knowledge brings an important perspective to the climate change discussion. Indigenous climate-related narratives can inform science (Alexander, Bynum, Johnson, King, Mustonen, Neofotis, Oettle, Rosenzweig, Sakakibara, Shadrin, Vicarelli, Waterhouse and Weeks 2011).

Adapting to Change

Indigenous Peoples are used to adapting to changes over long periods of time but climate change is happening drastically and abruptly (Trainor et al. 2007). The drastic pace of climate change has us testing the limits of our resilience. Nevertheless, the current changes lead Indigenous Peoples to strive to find ways to adapt and survive (Lewis 2008). In one such creative adaptation, some Native Alaskans are learning to grow vegetable gardens, given the increasing temperatures and lengthening growing season. This innovative response helps mitigate the high price of imported produce. Rather than simply a creative and happy solution, however, I think the Alaskan gardening project raises questions that so far have not received their due consideration. What are the implications of a hunter becoming a gardener? Given the strong links between indigenous means of subsistence and culture how might the changing food supply also change culture, and other aspects of life?

Indigenous Peoples are among the poorest and most marginalised people in the world. Poor and marginalised people have less ability and fewer resources to respond to climate change (Harries 2010). Just as settler societies have made the greatest contributions to climate change, so must they rise to the occasion and use their vast resources to address this problem which is largely of their making.

While Indigenous Peoples are more than willing to partner in developing and implementing thoughtful responses, we cannot shoulder this burden on our own.

Ironically, Indigenous Peoples face an additional burden. Although traditional peoples have done little to contribute to climate change, we often experience a backlash when Indigenous hunting and gathering is vilified in the context of reduced resources (Lewis 2008). Climate change must not be used as an excuse to vilify traditional practices.

Conclusion

While some level of change is inevitable, efforts to reduce the negative impacts of climate change must continue. To this end, partnerships between Indigenous and non-Indigenous Peoples must be strengthened and colonising societies must accept responsibility for the disproportionate burden placed on indigenous populations. Due to drastic changes in the natural environment, Indigenous Peoples are faced with adaptations that may alter their fundamental understandings of being indigenous. Peoples whose very existence is based on hunting and fishing the Native species of their traditional territories are faced with substantial changes when those species diminish or disappear. Some medicinal plants may cease to exist, thus necessitating changes in the way we preserve wellness and treat illness. If animals and plants necessary to our ceremonial practices are gone we cannot continue our cultures in ways that follow our original instructions. Drastic changes in the environment by definition mean drastic changes in the Peoples whose life and livelihoods are intimately connected to those environments. Environmental change means cultural change for Indigenous Peoples.

That being said, Indigenous Peoples have always demonstrated resilience and an ability to adapt to change. The drastic pace of environmental change is unprecedented so the way forward is unclear. My Lakota teachings tell me that seven generations ago my ancestors planned for me in ways that assured a place for me in this contemporary world. I have the same responsibility to plan for the next seven generations. In that way, we continue as Indigenous Peoples. While the future of our children is uncertain in that we don't know what it will look like, Indigenous Peoples will persist. *Mitakuye Oyasin.* All my relations.

References

Alexander, C., Bynum, N., Johnson, E., King, U., Mustonen, T., Neofotis, P., Oettle, N., Rosenzweig, C., Sakakibara, C., Shadrin, V., Vicarelli, M., Waterhouse, J. and Weeks, B. (2011) Linking indigenous and scientific knowledge of climate change. *BioScience, 61*(6), 477–484.

Furgal, C. and Sequin, J. (2006) Climate change, health, and vulnerability in Canadian northern aboriginal communities. *Environmental Health Perspectives,* 114(12), 1964–1970.

Goldtooth, T.B.K.(2010) Earth Mother, pinons, and apple pie. *Wicazo Sa Review, fall,* 11–28.

Harries, K. August, (2010) Health concerns relate to climate change in Canada's fourth world. *Windspeaker,* 20.

Hunter, E. (2009). Radical hope and rain: Climate change and the mental health of Indigenous residents of northern Australia. *Australasian Psychiatry,* 17(6), 445–452.

Lewis, S. January, (2008). Climate change a threat to indigenous cultures. *Windspeaker.* 12.

People's agreement (2011). Fact sheet retrieved 28th Jan 2014 from http://pwccc.wordpress.com/support/.

Thompson, I. February. (2010) Involve First Nations in combating climate change. *Windspeaker,* 8.

Trainor, S.F., Chapin, F.S., Huntington, H.P., Natcher, D.C. and Kofinas, G. (2007) Arctic climate impacts: Environmental injustice in Canada and the United States. *Local Environment, 12*(6), 627–643.

Weinhold, B. (2010) Climate change and health: A Native American perspective. *Environmental Health Perspectives, 118*(2), A64-A65.

Williams, J. (2012) The impact of climate change on indigenous people – The implications for the cultural, spiritual, economic and legal rights of indigenous people. *The International Journal of Human Rights, 16*(4), 648–688.

Is Situational Migration Vulnerable? An Answer from Asia through Sustainable Intervention

Sarangadhar Samal

In 1955 in the fishing village of Goradapal in India, a group of 32 fishermen and their families were displaced after a devastating flood of the Brahmani River. As resettlement was impossible, the entire village relocated to a B-class forest reserve adjacent to Raitala village in the Gondia block of Dhenkanal district, Odisha. Located at the foothills of Aswakhola, the settlers named their new habitat, 'Nayakateni'. 'Naya' means 'new' and 'kateni' means 'habitable location'. Unbeknownst to them, their relocation was an encroachment on reserve forest area which caused forest officials to take legal action by making arrests and issuing fines. In this new area, the lack of a water body prevented men from fishing and in effect, prevented women from selling fish at the market. The villagers were forced to abandon their fishing and boat crew occupations, which left most people on the brink of starvation. At that time, the area was arid, upland, and inadequate for cultivation and other farming activities due to lack of irrigation potential. In this situation, the settlers had no source of income for their survival.

The Nayakateni village area was surrounded by lush forest and the villagers cut down trees to sell as fuel to earn money to support themselves. As a result of many years of deforestation, the forest vanished resulting in the Nayakateni villagers having to travel long distances, to expend limited resources, and to spend more time in collecting wood and timber. For example, excluding the rainy season, female members of each household would walk four kilometres to a distant village to fetch potable water. Additionally, there was no schooling facility for the children, no electricity, no road leading to the village, and no real houses.

As the villagers continued to struggle, the National Youth Service Action & Social Development Research Institute (NYSASDRI) reached out to the Nayakateni village to provide support. With its head office only eight kilometres away, NYSASDRI interacted with villagers to create a micro-plan focused on socioeconomic and environmental development that would revitalise village operations and develop local natural resources through programmatic interventions. After conducting a community needs assessment NYSASDRI took various developmental initiatives to improve the lives of people in Nayakateni by building community capacity and by empowering people in the village to live sustainably. The first initiatives included the formation of community-based organisations such as a Women's Self-Help Group, an Adolescent Club, a Village Education Committee, and a Village Development Committee. The goal of NYSASDRI was ultimately to strengthen the community organisations and build their capacity for financial savings and credit. They introduced small entrepreneurship activities including but not limited to tailoring, animal husbandry, and vegetable cultivation. Leadership and enthusiasm began to flourish among the villagers. Further initiatives incorporated the construction of an open well which subsequently led to an organised campaign to teach awareness about safe drinking water. Additionally, a village school was established to help displaced children gain an education. A critical step in advancing village activities was the construction of a road that connected the village to mainstream development and which opened opportunities for using healthcare programs.

However, the above programs did not meet basic family requirements (food, clothes and shelters) so they became more dependent on harvesting wood, which caused rapid forest degradation in the area. As the village of Nayakateni did not have registered land rights or officially have permission to harvest wood, the forest department harassed them and forced villagers into paying bribes, which further led to socioeconomic strife.

In desperation, the Nayakateni people asked NYSASDRI to aid in the construction of a multipurpose irrigation dam that would support livelihoods in pisciculture (fish farming) and irrigation agriculture as well as satisfy their daily needs for water. NYSASDRI agreed to finance the construction of a multipurpose irrigation dam at the foothills of Swachala mountain. The dam took two years to complete, and NYSASDRI and Nayakateni residents went through many legal cases in the process: Nearby communities, political figures, and the forest department opposed the dam's progress. In the face of adversity the multipurpose irrigation dam was completed in 1990 and allowed villagers to begin new agricultural operations. Currently, the irrigation structure is supporting crops adjacent to the dam while the community uses water for pisciculture, for bathing, and for domestic animals' needs.

NYSASDRI provided instant support to Nayakateni through its various interventions that aimed at advancing the abilities of villagers to live their lives. Improvements to the village included introduction of irrigation agriculture, electricity, smokeless *chulhas* (firewood-saving stoves), fruit-bearing trees, and capital for micro-entrepreneurship. Furthermore, NYSASDRI contacted government officials for Nayakateni's land registration and eventually they were granted land rights. Each family was granted their own homestead with agricultural land. In effect, people relinquished most of their needs that previously could only be satisfied with forest resources. After the village became self-sustaining, the forest department, with help from NYSASDRI, formed a committee that enlisted a rotating group of Nayakateni villagers to monitor and protect the forest. From the time of migration to the present day, the number of households grew from 32 to 60. Each family has cultivation opportunities, contributes their fair share to preserving

Table 5.1 Timetable. Within 18 years changes occured through the direct guidance of NYSASDRI, and NYSASDRI handed over its responsibility and ownership to the villagers and withdrawn its activities. The villagers fought a long battle for their survival. This village was declared an ideal village by the government in 2008 and they are taking a leadership role in supporting many other villages

Year	**Situation of the villagers of Nayakateni**
1955	32 fisherman families of Goradapal were displaced after a devastating flood in Brahmani River – lost their shelter & occupation fishing, boating
1956	Illegally resettled at B class reserve forest & encroachment in reserve forest area
1957	Stealing & cutting forest resourses and sales fuel wood and timber at nearer market
1957–1985	Forest officials booked cases against them and the case was continue until 1985
1985	Forest was destroyed, disappeared and became barren of all area. there was no alternatives income source for their survivals as the area became dry, upland and inadequate for cultivation and farming activities due to lack of irrigation potentials

Year	Situation of the villagers of Nayakateni
1986	Villagers of Nayakateni requested to NYSASDRI for construction of a "MULTIPUPOSE IRRIGATION DAM" and NYSASDRI entry in the village
1987	Prepared micro-plan focused on socioeconomic and environmental development formation village level different committee and capacity building training provided to them, initiated micro entrepreneurship through financial support.
1988	Multipurpose Irrigation structure was started and forest department seized materials and equipment and booked case in the name of NYSASDRI and villagers (Forest land)
1989–1990	NYSASDRI taken initiative on livelihood support, capacity building training, financial support, education, drinking water, health care and vocational training
1991	Multipurpose Irrigation structure was completed and villagers started cultivation in the forest land
1992–2003	Government primary School, village road, Community centre established. But they did not have Land right of record
2004	They received RoR on Land and government recognise them as revenue village

the forest, and has the prospect of leading a fulfilling life. The remarkable history of the change of this village to sustainability can be followed in Table 5.1

Impacts

The social impacts on Nayakateni village have been profound and life-changing. The livelihood and survival of a fishing village was completely altered on their relocation which forced them to rely on agricultural practices. Furthermore, people benefited from the establishment of cohesive groups for self-help, education for children and illiterate women, and public distribution services. People began generating income from agricultural practices and utilisation of natural resources, which led to financial savings, credit possibilities and access to banks. Women began participating in village meetings and would send their children to school on a regular basis. Villagers learned new methods for cooking and learned the importance of boiling drinking water before consumption. Additional impacts relate to the agriculture development of 500 acres of wasteland, primarily through the provision of sufficient irrigation facilities. A works committee was established to handle water usage, to minimise conflicts regarding natural resources, to increase effective crop production, and to ensure food security.

From an ecological standpoint, arable areas have increased tremendously. Irrigation facilities have ensured one crop and in some areas, two crops. Villagers have adopted organic agricultural practices, including the use of biofertilizers, green manure, and other locally available materials. Additionally, there has been a positive change in soil moisture and an observed increase in groundwater levels in the open wells downstream. Also, crop production increased which led to an increase in the size of the village work-force. They have also protected local forest areas, which has created a thriving natural environment. The improvement in local awareness about the importance of green living led local community members to form a forest protection committee that repopulated the land with fruit-bearing trees and appointed a forest guard service to prevent future deforestation. The effects of NYSASDRI's intervention have been widespread. They helped create a considerable change in availability of food, living standards, accruement of household assets, literacy levels, and awareness of personal and environmental health.

Summing up, some of the qualitative changes and benefits to people and the environment over the years are listed in Table 5.2:

Table 5.2 Qualitative changes and benefits to people and the environment

1. 32 families became more than 60.	2. 60 Families received land registration from the government
3. Increased source of income for villagers	4. All families having houses, movable assets, and a bank account
5. All families maintaining their basic needs	6. Establishment of a government-run primary school and an integrated child development centre has increased literacy in the village
7. Improvements in infrastructure: electricity in the village, construction of a reliable road, pucca homes for each family and installation of a hand pump	8. The village is hosting biannual festivals
9. Selected and appointed an Accredited Social Health Activist (ASHA)	10. Increased participation of women at Pallisabha and Gramsabha (village and Grampanchayat-level meeting)
11. Increased women participation in decision making and leadership processes	12. Creation of a forest protection service
13. They have participated and represented Nayakateni in Panchayati Raj System	14. 4 Women groups and 2 male groups are practicing savings and credit programs.
15. Village development committee working in the village	

Chapter 6
Integrating Economic, Environmental and Social Perspectives

Economic Development, Environmental Development and Social Protection are the three fundamental "pillars" for the integration of sustainable development, according to UNCSD 2012 (United Conference on Sustainable Development). The UN Rio summit (Rio +20) was celebrating the 20th anniversary of the first UN climate summit establishing Agenda 21, held in Rio de Janeiro 1992.

The Social Protection Floors Recommendation R. 202 (2012) Adopted by the International Labour Organization

Alejandro Bonilla Garcia

At its 304th session, the Governing Body of the International Labour Office decided the second recurrent discussion at the 100th Session of the International Labour Conference in 2011 would be devoted to the strategic objective of social protection. For such recurrent discussion, the ILO prepared a report highlighting the key elements that would be relevant for a discussion on a possible Recommendation. The report of the ILO cognised and reiterated that social security is a human right as stated in Article 22 of the Universal Declaration of Human Rights; that it is also a human necessity since it provides income security, prevents and reduces poverty and inequality, and promotes inclusion and dignity; and finally that it is also an economic necessity since it constitutes an automatic stabiliser in times of crisis, mitigates the economic and social impact economic downturns, and enhances resilience.

After thorough and lengthy preparation and discussions in a Tripartite Committee, the International Labour Conference adopted at its 101th Session, Recommendation R.202 (2012) concerning national floors of social protection. For memory, it is important to recall that an ILO Recommendation is an international labour standard; it is tripartite and universal and that all ILO Member States are obliged to consider its application. An ILO Recommendation cannot be ratified and does not create binding obligations for Member States, but they still have the obligation to submit it to their competent authority for giving effect to it by national legislation and other measures. Member States have to inform the ILO DG on the measures taken to bring the Recommendation to the competent authorities and are obliged to report regularly, at the request of the Governing Body of the ILO, on its application in law and practice.

Recommendation 202 on nationally defined social protection floors, has five sections. The first is a preamble where it recognised that social security is a social right and an economic necessity. Section I of the recommendation defines the agreed objectives scopes and principles. Section II defines the elements of a social protection floor. Section III addresses the issues related to the extension of social security and Section IV focuses on supervision and monitoring issues.

Recommendation 202 is clear in stating that social protection floors are a fundamental element of national social security systems and that they should be established and maintained within strategies for the extension of social security that progressively ensure higher levels of social security to as many people as possible, guided by ILO social security standards. According to R. 202, national strategies for the extension of social security should prioritise the implementation of social protection floors as a starting point for countries that do not have a minimum level of social security guarantees as a fundamental element of their national social security systems; and seek to provide higher levels of protection to as many people as possible reflecting economic and fiscal capacity.

For this purpose, ILO Member States agreed to progressively build and maintain comprehensive and adequate social security systems compatible with national policy objectives and seek to coordinate social security policies with other social policies. Progressive realisation would consider: a) setting targets and timeframes; b) coherence with social, economic and employment policies; c) regular monitoring of implementation and periodic evaluation; and d) tripartite participation and consultation with representatives of persons concerned.

R. 202 defines a set of basic social security guarantees which secure protection aimed at preventing or alleviating poverty, vulnerability and social exclusion. The guarantees should ensure at minimum that, over the life-cycle, all in need have access to essential health care and to basic income security which together secure effective access to goods and services defined as necessary at the national level. Also R.202 establishes that national social protection floors should consider at least the following four guarantees: 1) access to a set of goods and services constituting essential health care including maternity care; 2) basic income security for children; 3) basic income security for persons in active age unable to earn sufficient income and 4) basic income security for persons in old age.

One of the most important elements of R. 202 is that it states that the above-mentioned guarantees should be provided to at least all residents and children, as defined in national laws and regulations, subject to Members' existing international obligations, on the following principles: universality of protection based on solidarity; social inclusion considering persons in the informal economy; non-discrimination, gender equality and responsiveness to special needs.

Recommendation 202 establishes that basic social security guarantees should be established by law and that relevant national laws and regulations should: a) specify the range, qualifying conditions and levels of the benefits giving effect to these guarantees; b) be impartial, transparent, effective, simple, rapid, accessible and inexpensive complaint and appeal procedures should also be specified; c) consider complaint and appeal procedures free of charge to the applicant.

Recommendation 202 reaffirms that social security and social protection, including social protection floors, are an overall and primary responsibility of the State; that benefits should be adequate and predictable and based on non-discrimination, gender equality and responsiveness to special needs. Finally R. 202 considers that the access to benefits should respect the rights and dignity of people covered as well as respect for collective bargaining, freedom of association and social dialogue.

As regards financing and delivering, R. 202 indicates that countries should consider using a variety of different methods to mobilise the necessary resources to ensure financial, fiscal and economic sustainability of national social protection floors. National social protection floors should be financed by national resources. Countries whose economic and fiscal capacities are insufficient to implement the guarantees may seek international cooperation and support that complement their own efforts. While considering a diversity of approaches, solidarity in financing and fair balance of interests should be considered. Transparent, accountable and sound financial management and administration are required. Financing should consider: a) financial, fiscal and economic sustainability with due regard to social justice and equity, b) high-quality public services that enhance delivery of social security systems and c) coherence across institutions responsible for the delivery of social protection.

Recommendation 202 indicates that national definitions of minimum levels of protection should meet a number of requirements. The first is that persons in need of health care should not face hardship and an increased risk of poverty due to the financial consequences of accessing essential health care. The second is that free pre-natal and post-natal medical care for the most vulnerable should also be considered. The third is that basic income security should allow life in dignity and possible reference points are national poverty lines or income thresholds for social assistance. R. 202 indicates that levels of basic social security guarantees should be regularly reviewed through a transparent procedure that is established by national laws, regulations or practice, with ensured tripartite participation with representative organisations of employers and workers, as well as consultation with other relevant and representative organisations of persons concerned.

Recommendation 202 indicates that countries should aim to achieve the range and levels of benefits set out in Social Security (Minimum Standards) Convention, 1952 (No. 102), or in other ILO social security Conventions and Recommendations setting out more advanced standards. Countries should consider ratifying, as early as national circumstances allow, the Social Security (Minimum Standards) Convention, 1952 (No. 102). Furthermore, countries should consider ratifying, or giving effect to, as applicable, other ILO social security Conventions and Recommendations setting out more advanced standards. R. 202 complements existing ILO social security standards, and it does not lower these standards.

As regards nationally-defined monitoring mechanisms and national consultations, R. 202 establishes that countries should regularly convene national consultations to assess progress and discuss policies for the further horizontal and vertical extension of social security. Also that strategies and mechanisms should be established for the collection, compilation and publication of adequately disaggregated social security data, statistics and indicators as well as the legal framework to protect private individual information. All of the above, is taking into consideration guidance provided by Resolution of the International Conference of Labour Statisticians and other statistical standards exchange of information, experience and expertise among members and with ILO technical assistance from ILO and others.

ILO's two-dimensional strategy for the extension of social security is considered by R. 202: The Vertical dimension considers that countries should progressively ensure higher levels of protection, guided by Convention No.102 and more advanced standards; the Horizontal dimension considers that countries should guarantee access to essential healthcare and minimum income security for all.

Currently the ILO and the World Bank are co-chairing the Inter-Agency coordination board (SPIAC-B) which includes UN agencies and donor countries to coordinate activities at the global and national level as well as to develop instruments to identify needs, possibilities and paces to gradually implement social protection floors world-wide.[1]

Inclusion of Children as Stakeholders in Social, Economic and Environmental Development

Antoinette Lombard and Andries Viviers

Introduction

In 2011, during the 17th Conference of the Parties to the United Nations Framework on Climate Change (COP17) in South Africa, particular attention was drawn to the widespread impact of climate change on vulnerable groups, especially children, who are likely to bear the brunt of climate change although they contributed least to the problem (Reid, Huq and Murray 2010, UNICEF 2011a, United Nations 2012). The impact of climate change on children in South Africa should be considered in relation to wider development pressures affecting the country, including high levels of poverty and inequality, the effects of HIV and AIDS, the management of scarce natural resources, rapid urbanisation and gender inequality (Pettengell 2010, UNICEF 2011b).

1 For more and continuously updated information on the implementation of social protection floors world-wide please visit the following address: http://www.social-protection.org/gimi/gess/ShowMainPage.do?ctx=0.

The importance of including children in global environment and development is set out in Agenda 21 (Chapter 25) adopted at the Earth Summit in Rio de Janeiro in 1992 (UNICEF 1992). It is supported in *Principle 3* of the 1992 *Rio Declaration on Environment and Development*, which states that the 'right to development must be fulfilled so as to equitably meet developmental and environmental needs of present and future generations' (United Nations 1992: 1). Globally, climate change poses a severe threat to human development (Veton n.d.). It has had and will continue to have a profound impact on children's right to survival, development, protection and participation (Urbano and Maclellan 2010).

The Republic of South Africa (RSA) has already taken steps to understand, recognise and address the challenges posed by climate change, as is evident in its National Climate Change Response strategies and other environmental and developmental policy responses (Department of Environmental Affairs 2011, RSA, 2012, UNICEF 2011b). However, despite these strategies, generally, 'children remain invisible because the majority of South Africa's climate change policies and programmes, whether they be at national, provincial or district levels, do not yet adequately recognise children's vulnerabilities, specific needs and the role children can play as agents of change at the grassroots level' (UNICEF 2011b: 9).

In recent years, children have demonstrated that they are active partners in the response to climate change, and that their participation in local and global debates on the matter is important (Burgess 2012, Children in a Changing Climate 2008, United Nations 2009). This contribution supports this movement and shows some South African children's views on the impact of climate change and on how they can take responsibility for protecting the environment by integrating social and economic development. The discussion then moves to children's inclusion as stakeholders of sustainable development, concluding that children can play a role in limiting their increased risk and pre-existing vulnerability relating to the impact of climate change on their environment, and social and economic development.

Children's Views on the Impact of Climate Change

A joint qualitative study undertaken by UNICEF and the South African Department of Women, Children and People with Disabilities, and the Department of Environmental Affairs, was reported as *Exploring the Impact of Climate Change on Children in South Africa* (UNICEF 2011a). It was complemented by the study *Change through the Eyes of a Child: South African Children Speak about Climate Change* (UNICEF 2011b). These studies indicate that children have a comprehensive understanding of the interrelation between climate change and the environment, and how change affects scarce resources, poverty, food insecurity, health, education, infrastructure, the economy and social and emotional well-being. Children argue that adults have contributed to climate change, but do not blame them for this: 'Not because they [adults] have bad hearts or anything like that, but because they do not have the information' (UNICEF 2011a: 11).

Children can identify roles and responsibilities for themselves, families, schools, the broader community and government in responding to and reducing the impact of climate change. These approaches include education campaigns and avenues to address poverty and earn an income, for example, by actions such as the following: ' ... the recycling team from the school takes all the litter to the recycling centre and then they get some money. With that money they help the poor community around our school' (UNICEF 2011a: 10). Another respondent explained:

> I started this project in my community whereby we pick up waste materials, almost each and every Saturday and then it's called 'greening our ways of greening'. It means that we encouraged people to go green and live green. By doing so we are educating them to reduce the amount of electricity that they use every day and to also, when they collect waste materials, to put them in a plastic bag. (UNICEF 2011a: 9)

Another suggested that government should 'fund projects of cleaning the environment' (UNICEF 2011a: 14).

Children do not only see themselves being affected by climate change, but as:

> the future generation; we are the people who this global warming is going to affect. (...) We are the future scientists; we are the future doctors who are going to have to deal with diseases and our lives because of global warming. We are the people who are going to suffer; we are the economy that is going to suffer (UNICEF 2011a: 5)

Their desire to be included and belief that they have the capacity to contribute is evident: ' ... if anything is to happen you go past them first, so I think our voices should be heard because we are the future' (UNICEF 2011a: 11).

Children's Inclusion as Stakeholders in Sustainable Development

Recent reports by the Institute for Development Studies [IDS] (2009, 2012), Save the Children (2010), Strazdins and Skeat (2011) and UNICEF (2011a) on children's participation in the climate change arena, which encompass social, economic and environmental development, indicate that children

- are able to contribute to sustainable development, the mitigation of climate change and adaption strategies related to these;
- bring unique and complementary insights to these development issues that can enrich strategies;
- can take an equal place, alongside other stakeholders, in claiming their right to be informed and participate in social, economic and environmental development.

Recognising children as important stakeholders in sustainable development implies advocating and lobbying for their inclusion and participation in decisions on climate change and broader development. Burgess (2012: 8) argues that 'children can be key actors in driving the sustainability agenda forward, both now and in the future as adults, so they should be included and consulted on sustainability initiatives at the earliest possible moment'.

In the context of climate change, food insecurity and poverty are critical to sustainable development. Increased poverty rates and disparities are of the utmost concern for social workers (Kaiser 2011, Skoufias, Rabassa and Olivieri 2011), especially because they increase children's extreme vulnerability and risk. Social workers are well positioned to advocate for the right of children to be heard. Advocating and lobbying strategies for child participation include a human rights approach which highlights equality and equity. UNICEF (2008: 6) argues that inequity across countries can 'easily be extended to inequities between adults and children', which raises the issue of intergenerational justice implicit in sustainable development. As human rights professionals (Ife 2012), social workers are key players in ensuring that children claim their rights to participate in development and the implementation of policies.

With the adoption of a developmental social welfare policy for South Africa (RSA 1997), social work has been mandated to harness social development as an approach to promote the wellbeing of the population in conjunction with a dynamic process of economic development (Midgley 1995), and hence to adopt human rights approaches to achieve integrated human, social and economic development (Lombard 2007).

Conclusion

Children are often considered in terms of inherent vulnerability, but it is a mistake to think of them as 'passive bystanders' or 'helpless victims' in the climate change discourse and agenda (Children in a Changing Climate 2008: 3). In future, children will be the ones most affected by climate change, so they should have a voice today, and their views and insights must be taken into account (Children in a Changing Climate 2008). Children are willing and able to participate actively in decisions that affect their lives (UNICEF 2011a). They understand their role as the generation that has to prevent further climate change, and has to find sustainable solutions which will create hope for future generations. Policy and decision-makers should listen to the voices of children, because children have a right to participate, and because listening to their voices is an important component of climate justice.

This paper shows that there is growing evidence of the positive role children can play in mitigating climate change. However, society (especially adults) has to reconstruct its view of children as active citizens in social, economic and environmental development. Children may require additional supportive measures or spaces where they can be more active agents of change (UNICEF 2011a) and integrate their rights to protection, survival, development and participation. The inclusion of children lies at the heart of social, economic and environmental justice, and social workers have an obligation to facilitate this inclusion to ensure sustainable development in the interests of current and future generations.

References

Burgess, J. (2012) *Rio+20: Making a Better World for Children*. London: UNICEF UK.

Children in a Changing Climate. (2008) *A Right to Participate: Securing Children's Role in Climate Change Adaptation*. London: Children in a Changing Climate.

Department of Environmental Affairs. (2011) *National Climate Change Response White Paper.* Pretoria: Department of Environmental Affairs.

Ife, J. (2012) *Human Rights and Social Work. Towards Rights-based Practice.* 3rd edition. London: Cambridge University Press.

Institute for Development Studies (IDS). (2009) In *Focus Policy Briefing 23: Children, Climate Change and Disasters*. Brighton: Institute for Development Studies.

Kaiser M.L. (2011) Food security: an ecological-social analysis to promote social development. *Journal of Community Practice*, 19(1), 62–79.

Lombard, A. (2007) The Impact of Social Welfare Policies on Social Development in South Africa: An NGO Perspective. *Social Work/Maatskaplike Werk* 43(4), 295–316.

Midgley, J. (1995) *Social Development. The Developmental Perspective in Social Welfare*. London: Sage.

Pettengell, C. (2010) *Climate Change Adaptation: Enabling People Living in Poverty to Adapt*. London: Oxfam GB.

Reid, H., Huq, S. and Murray, L. (2010) *Community Champions: Adapting to Climate Change Challenges*. London: International Institute for Environment and Development.

Republic of South Africa (RSA). (1997) Ministry for Welfare and Population Development. White Paper for Social Welfare. Notice 1108 of 1997, *Government Gazette*, 386(18166). Pretoria: Government Printer.

Republic of South Africa (RSA). (2012) National Planning Commission. *National Development Plan: Vision for 2030*. Pretoria: Presidency.

Save the Children. (2010) *Right to a Future: Climate Change Negotiations Must Be Accountable to Children*. London: Save the Children.

Skoufias, E., Rabassa, M. and Olivieri, S. (2011) *The Poverty Impacts of Climate Change: A Review of the Evidence*. Washington, DC: World Bank.

Strazdins, L. and Skeat, H. (2011) *Weathering the Future: Climate Change, Children and Young People, and Decision Making*. Canberra: Australian Research Alliance for Children and Youth.

UNICEF. (1992) *Children and Agenda 21: A Guide to UNICEF Issues in the Earth Summit's Blueprint for Development and Environment into the 21st Century.* Geneva: UNICEF.
UNICEF. (2008) *Climate Change and Children: A Human Security Challenge.* Florence: UNICEF Innocenti Research Centre.
UNICEF. (2011a) *Change through the Eyes of a Child: South African Children Speak about Climate Change.* Pretoria: UNICEF South Africa.
UNICEF. (2011b) *Exploring the Impact of Climate Change on Children in South Africa.* Pretoria: UNICEF South Africa.
United Nations. (1992) *Rio Declaration on Environment and Development.* New York: United Nations.
United Nations. (2009) *Growing Together in a Changing Climate: The United Nations, Young People and Climate Change.* New York: United Nations Joint Framework Initiative on Children, Youth and Climate Change.
United Nations. (2012) *Report of the Second Committee: Sustainable Development: Protection of Global Climate for Present and Future Generations of Humankind A/67/437/Add.4.* New York: United Nations.
Urbano, M. and Maclellan, N. (2010) *Climate Change and Children in the Pacific Islands.* Suva: UNICEF Pacific.
Veton, C.C. n.d. *The Benefits of a Child-centred Approach to Climate Change Adaptation.* London: UNICEF UK & Plan International.

Multidisciplinary Education for Environmental Sustainability

Cathryne Schmitz and Thomas Matyók

The World Commission on Environment and Development (1987) recognised the interconnection across peace, security, development, and environmental sustainability. Likewise, war and extreme economic inequality are linked to structural violence in the human community and the degradation of the biophysical environment. Because the issues are interconnected, creating communities supportive of environmental sustainability is a complex, multi-dimensional process, which is linked to concerns for positive peace and social, economic, and political justice (Schmitz, Matyók, Sloan, and James 2012).

Beyond Disciplinary Response

The immense and multifaceted nature of the concerns cuts across disciplines and consequently demands responses that are interdisciplinary and inclusive of the social and natural sciences. While the natural sciences have an established body of knowledge in environmental studies, the human sciences lag behind. The human science fields, particularly social work and peace and conflict studies (PACS), as professions committed to social and economic justice, community and team building, and collaboration, are uniquely situated to provide leadership in creating multidisciplinary processes of engagement in the work toward the development of environmentally sustainable communities.

Professional disciplines, however, too often educate students within silos leaving graduates ill-prepared to join multidisciplinary efforts in response to an environmental crisis (Orr 2011). Given the inadequacies of the current educational structure that values compartmentalised learning over cross-discipline collaboration, how do we educate students to engage as allies with other professional and community participants in a change process toward environmental sustainability? The design disciplines, which are transformative in nature, provide a guide toward the development of an inclusive model. While the scientific disciplines can tell us what something is and the humanities can tell us what it means, they fail to propel us toward the active creation of what is not yet present (Mendoza and Matyók 2013). In modelling the learning process after the transformative and inclusive nature of design disciplines, the potential exists for a co-creating a dynamic, flexible learning environment. New centres can be created on the margins, shifting us away from the construction of political, social, and economic elites that benefit from exploitation of the environment.

Another barrier to educating students about inclusive environmental practice exists within an academic context that does not integrate a focus on environmental concerns. In compartmentalised learning, the mega context for multidisciplinary analysis and learning does not come into focus. Consequently, courses focused on a narrow slice of environmental justice and sustainability exist in secluded pockets across campus without connection to each other. Further, disciplines often discount the language of other professions framing these "other" courses as unworthy of their focus. Students are neither aware of the range of courses nor of the connections across the issues. Setting the context at the mega level would provide the framework for cross-disciplinary connections linking fields of study, faculty, and students. A multidisciplinary network of faculty could create a model for motivating students.

Within a context where social workers are being called upon to enter the field of environmental practice (Coates 2003, Mary 2008) and PACS workers are engaging in the positive transformation of environmental conflict through sustainable peace building practices, an opportunity existed at the University of North Carolina Greensboro to create an interdisciplinary course focused on the multi-dimensional issues of environmental justice and sustainability. A faculty member from business, who focuses their work on social and environmental responsibility, joined in the process. In framing the approach to curriculum, social cubism (Byrne and Carter 2000) provided a model for the integration of horizontal and vertical thinking with multi-dimensional complexity. The social cube captures six dimensions of analysis that are necessary when investigating conflict: history, demographics, economics, religion, politics, and psychocultural factors. The cube can also help conceptualise disciplinary integration of the biological, hard, and technological sciences with the economic, political, and human sciences supporting analysis from multiple perspectives.

Reflection on Curriculum Development

A course was developed in which students work at the micro-, meso-, and macro-levels through the lens of mega analysis. At the core of the educational process is multidisciplinary team building in which students have a collective responsibility for the learning. Through this process students are immersed in a culture that values civic engagement and global citizenship. Within this collective setting students join with others as allies rather than rescuers and/or experts. These teams take on the dual role as learner and educator with responsibility for creating learning opportunities for other teams in the class.

Important to discuss here is the necessity for this type of classroom and learning to include interdisciplinary teams of faculty. In the classroom, faculty models the interdisciplinary cooperation that students are being asked to manifest in the field. The "teaching" is not simply a matter of dividing the available weeks and assigning responsibility to individual faculty for those time periods. Faculty co-teaches in real-time. Interdisciplinary faculties are always present to expand the learning.

The human systems (social, political, and economic) are actors on and within the ecological environment (see Schmitz, Matyók, James and Sloan 2013 for the model). It is through this lens that participants in the class are exposed to the impact of war, poverty, and human wellbeing within an ecological context. In addition to the history, global politics, and science of environmental concerns, students explore community development (Gamble and Weil 2010); alternative models of economics (Schumacher 1989, Shiva 2005); visionary models for environmental change (Coates 2003, Hawken 2010, Orr 2011); and conflict transformation toward the development of positive peace (Schmitz et al. 2012). The process of change is one in which the local community is recognised as the centre of expertise and knowledge building.

Structuring the Process

Multidisciplinary teamwork is integral to the development of the skills necessary to learning a process of co-creation. Within their team students discuss assigned materials, which include academic as well as popular books, articles, movies, and websites. They become educators for the class on an environmental issue, and explore an area of concern then co-create a response which is presented to the class. They are encouraged to be creative in their presentation in order to develop skills that can facilitate effective communication and change. The teams come to share beyond the constraints of disciplinary language, finding ways to work within community without elitist dialogue.

Students are introduced, often for the first time, to the need for cross-disciplinary work to address the *wicked problems*. These are problems, such as environmental sustainability, that have complex interdependencies making them "impossible" to resolve. Within the complexity, however, exists the opportunity for transformation. A case study method of teaching established a framework for the vertical and horizontal integration of learning required in addressing the *wicked problems* of environmental degradation and sustainability. A grid is used to structure learning at the mega-, macro-, meso-, and micro-levels (see Schmitz et al. 2013 for the grid).

Students analyse the issues then engage in team development of a community change project. These requirements contribute to a) the construction of citizenship by requiring diverse student groups to work together in addressing problems, and b) the development of community approaches to environmental sustainability by *chunking* larger issues. For instance, in exploring the impact of sustainable gardens developed by refugees on the reinvigoration of decaying communities, the team becomes engaged with local, national, and global food policy; the impact of pesticides; the misuse of water; and the economics of poverty and globalised mega farms. Student teams might also start with an interest in water then move to exploration of the politics of coastal development, fishing rights and the impact of over fishing, indigenous rights, water use and water rights, and the politics of draught and war. Or, a team may begin with an interest in solar energy then explore the local, national, and international economics and politics of resource development and the links to poverty, war, indigenous rights/practices, and the depletion of resources. Other teams have explored habitat loss and the connection to community and environmental degradation. All of these issues must be explored across the micro-, meso-, macro-, and mega-levels.

Lessons Learned

Experience has helped us create a learning environment that supports students in exploring issues related to environmental degradation, disrupting this process, and gathering the skills and knowledge for developing communities that support a healthy biophysical environmental. The wicked problems of our times, of which environmental sustainability is a leading one, demand expansive thinking that integrates knowledge and challenges disciplinary prejudices, which result in bounded and shallow thinking that privileges one way of knowing over all others.

References

Carter, N. and Byrne, S. (2000) The dynamics of social cubism: A view from Northern Ireland and Quebec, in *Reconcilable Differences: Turning Points in Ethnopolitical Conflict*, edited by S. Byrne and C. Irvin. West Hartford, CT: Kumarian Press, 41–64.

Coates, J. (2003) *Ecology and Social Work: Toward a New Paradigm*. Halifax: Fernwood Publishing.

Gamble, D.N. and Weil, M. (2010) *Community Practice Skills: Local to Global Perspectives*. New York: Columbia University Press.

Hawken, P. (2010) *The Ecology of Commerce: A Declaration of Sustainability* (revised edition). New York: Harper.

Lysack, M. (2010) Environmental decline and climate change: Fostering social and environmental justice on a warming planet, in *Transnational Social Work Practice,* edited by N.J. Negi and R. Furman. New York: Columbia University Press, 52–75.

Mary, N.L. (2008) *Social Work in a Sustainable World.* Chicago: Lyceum.

Mendoza, H.R. and Matyók, T. (2013) Designing student citizenship: Internationalized education in transformative disciplines. *The International Journal of Art & Design Education,* 32(2), 215–225.

Orr, D.W. (2011) *Down to the Wire: Confronting Climate Collapse*. Washington, DC: Island Press.

Schmitz, C.L., Matyók, T., James, C.D., and Sloan, L. (2013) Environmental sustainability: Educating on the environment in contemporary curricula in the USA, in *Environmental Social Work*, edited by M. Gray, J. Coates and T. Hetherington. New York: Routledge, 260–279.

Schmitz, C.L., Matyók, T., Sloan, L., and James, C.D. (2012) The relationship between social work and environmental sustainability: Implications for interdisciplinary practice. *International Journal of Social Welfare,* 21(3), 278–286.

Schmitz, C.L., Stinson, C.H., and James, C.D. (2010) Reclaiming community: Multidisciplinary approaches to environmental sustainability. *Critical Social Work,* 11(3), 83–95.

Schumacher, E.F. (1989) *Small is Beautiful: Economics as if People Mattered.* New York: Harper & Row.

Shiva, V. (2005) *Earth Democracy: Justice, Sustainability, and Peace.* Cambridge, MA: South End Press.

World Commission on Environment and Development. (1987) *Our Common Future*. New York: Oxford University Press.

Zapf, M.K. (2009) *Social Work and the Environment: Understanding People and Place*. Ontario: Canadian Scholars' Press.

Chapter 7
Social Action for Clean Water

The UN General Assembly declared on 28 July 2010 that safe and clean drinking water and sanitation is a human right essential to the full enjoyment of life and all other human rights.[1] What will be the role of social work and social policy to respect, protect and fulfill the right to clean water?

The Quest for Clean Water in Kenya: Social Workers' Uneasy Role

Pius Mutuku Mutie

Introduction

The number of publications on water is enormous and this is not surprising considering the many dimensions in which one could assess water. While some scholars focus on hydrology, others look at properties or quality, access, supply/distribution, irrigation, water as a commodity, sanitation, and climate change or water conflicts. There are others, particularly sociologists and anthropologists, who feel that there is a lot of emphasis on the economic value of water (commoditisation) ignoring the aspect of water as a social good or the water users' (communities') sentimental value for water (Day 1996). By and large, this chapter addresses the issue of access to clean water in a developing country and the role of social workers in this process.

Recent reports indicate that about 780 million people around the world lack access to clean and safe drinking water (WHO/UNICEF 2012). Needless to say, nearly all these people are in the developing world, with the majority in Africa. The report indicates that only about 61 per cent of the people in Sub-Saharan Africa have access to improved water supply sources, compared with about 90 per cent in other developing countries in Latin America, the Caribbean, northern Africa, and many parts of Asia. Inclusion of access to clean water in the Millennium Development Goals shows the significance with which the global community accords sanitation and hygiene in the developing world.

In spite of the fact that the goal of halving the proportion of the world's population without access to clean water by 2015 was achieved in 2010 (WHO/

1 (General Assembly declares access to clean water and sanitation is a human right) [consulted in May 2013].

UNICEF 2012), many rural populations in sub-Saharan Africa still face a water problem. The report shows that many of these African countries (Kenya included) are not on track to meet the MDG goal of clean water by 2015. Highlighted also were the 'stark disparities' between urban and rural populations when it comes to access to clean water. One of the reasons why access to clean water attracts global attention is the impact it has on other aspects of development. The Congressional Research Service (2012) shows that there is a link between the time spend looking for water or use of unsafe water and poverty. In any case, it is difficult to imagine how countries can eradicate extreme poverty and hunger or ensure environmental sustainability without addressing water needs.

One outstanding omission in most of the reports that provide critical data on the living conditions of world populations is the process through which a country has attained or failed to attain service delivery. Even where the reasons are provided, for instance, poverty, poor governance or corruption, the main players at the point of project implementation are often ignored. And this is partly what this contribution is about. Among other things, it seeks to discuss the status and process of water supply in Kenya and the role that social workers play or could play.

Kenya's Water Situation

According to Kenya's Vision 2030 (2008), the government's development blueprint, only 57 per cent of households in Kenya have access to water that is 'considered safe'. And the proportion is even lower (only 50%) for rural populations. These statistics are corroborated by other sources. The Congressional Research Service (2012) report, for instance, indicates that on average, only 59 per cent of Kenya's population has access to what it calls 'improved drinking water sources'. One notable thing is the disparity between urban and rural populations. Considering that 67 per cent of Kenya's population is rural (Kenya 2010), the fact that 50 per cent of this population (about 15 million people) has no access to clean water demonstrates the enormity of the problem. Besides, there are many incidences of water-borne diseases such as Bilharzia.

Equally important to highlight is that, obscured in these statistics, are the long distances some community members have to cover to get water that is 'safe'. There are many households in Kenya whose nearest water source (a river, stream, well or a dam) is more than 10 kilometres away. So the priority in many arid and semi-arid parts of Kenya has actually been one of access to 'just' water and not necessarily 'clean' water. It is prudent here to clarify that only 20 per cent of Kenya's landmass is considered 'wet' (receives adequate rainfall) with the rest (80%) being either arid or semi-arid lands (ASAL). The ASAL, very important for livestock-keeping and supply of beef, not only experience recurring droughts and acute water shortages but also famines. These areas, mainly in the northern and eastern parts of the country, have also been marginalised in terms of national resource distribution. It is against this background that the Kenya Constitution 2010 made a provision for an 'equalization fund' so that areas that have lagged behind can be uplifted. The fund is expected among other priorities, to assist in the development of sustainable water projects.

Initiatives to Improve Access to Water in Kenya

In an article that traces water and sanitation development in Kenya from 1895 to 2002, Nyanchaga and Ombongi (2007) document some of the earliest initiatives in water supply. Having occupied a largely dry colony compared to Uganda for instance, Britain's Imperial British East Africa (IBEA) Company started water programmes in what was later to become Kenya. This was done as a matter of necessity; to improve the living conditions of white settlers and to some extent, Africans. The authors note that the first piped water supplies to towns such as Nairobi and Mombasa were run by what was then called the Uganda Railway which connected Mombasa at the coast to Kampala in the interior.[2]

By 1914, they note further, private investors had ventured into the public water sector. Other developments in the water sector were contained in the Swynnerton Plan of 1954 which stipulated various ways of increasing water supply not only to urban but also rural areas. The authors note that after Kenya's independence in 1963, water was equally given priority by the new African administration. And just like in the colonial period, it appears that external interventions continued to play a major role. Of significance were two multilateral organisations, the World Health Organization (WHO) and United Nations Children's Fund (UNICEF) which started an initiative just before independence in August 1960 dubbed 'Environmental Sanitation Programme'. The initiative saw many rural communities gain access to clean water. The Swedish government is cited as having signed a credit agreement with the Kenya Government in 1970 to assist also in rural water supply and perhaps to support the earlier initiative of the WHO and UNICEF.

While it is clear that access to water continued to improve all the way to 2002, some conspicuous issues in Nyanchaga and Ombongi's article include the multiplicity of players in the water sector over the years, the lopsided attention given to urban water supply and overlapping of legislations. There is also very little discussion, if any, on the roles of communities or community mobilisers (e.g. social workers) in the struggle to access water and in the sustenance of water projects.

Problems facing the water sector, ranging from administrative and bureaucratic issues to sustainability of water projects, led to the enactment of the Water Act of 2002 (Kenya 2002). The Act established a number of bodies while specifying the roles of existing ones. The bodies that were created and their roles specified in the Act are Water Reform Programme, Water Resources Management Authority, Catchment Area Advisory Committee, Water Services Regulatory Board, Water Services Boards, Water Services Providers, Water Services Trust Fund and Water

2 Now Kenya Railways.

Appeals Board (Kenya 2002, KWAHO 2009). By law, every water resource in Kenya is vested in the State and water resources are managed by the Water Resources Management Authority.

Looking at the delineated functions of these entities, there is hardly any mention of communities who are the eventual water users. One gets the impression that the water sector, even after the historical reforms realised through the 2002 Act of parliament, still adopts a top-down approach. It is only the Water Service Boards where there is mention of a direct link with water users through the community based organisations (CBOs). Among other things, the boards should assist CBOs to access funds from the Water Services Trust Fund. But just like in the historical account given above, one notices that social workers do not feature in the Water Act as contributors to the development of the water sector. Although the Act is said to have been developed after broad consultations with 'stakeholders', it is clear that social workers or indeed community development workers were not involved in this initiative or their views were ignored.

Players in the Water Sector

In a country where only half of the rural population has access to clean and safe water, water supply has preoccupied individuals, groups, communities and institutions for a long time. The responsibility of water provision to rural populations largely rests with the communities concerned. Many of these communities get water from rivers, streams, wells and springs. Even in cases where the government or non-state actors have funded water projects such as roof catchments, dams or boreholes, or spring protection, communities bear the responsibility of operation and maintenance once the external support is withdrawn.

The other important player in the water sector is indeed the Kenya government. The water ministry, whose name varies with time (previously called Ministry of Water and Irrigation but changed in 2013 to Environment, Water and Natural Resources), has been one of the most important dockets in cabinet. And this significance is not just because water supply is critical in Kenya but also due the sheer number of development organisations that fund the water sector. These include multilateral organisations such as the World Bank, UNICEF and WHO; bilateral agencies like Swedish International Development Agency (SIDA), Norwegian Agency for Development (NORAD), Danish International Development Agency (DANIDA), Department for International Development (DFID), Japanese International Cooperation Agency (JICA) and international non-governmental organisations such as Action Aid, Water.org, Oxfam and CARE among others. As shown earlier, some of these agencies have been involved in the water sector for a long time. And it is in some of these non-state actors where social workers play at least some role in water service delivery. Otherwise, the state's water ministry does not employ social workers.

Social Work and 'Industry'

From a theoretical perspective, there has always been contrasting schools of thought when it comes to 'social work and industry'. Some feel that social workers working in industries deal with only the welfare aspects of the organisation and not the core business of that particular enterprise. Many organisations do not see social workers' roles beyond the issue of welfare 'if at all' (Kumar 2006). It is has been shown that social workers who play 'helping roles' in an organisation, are given lower status compared to management professionals (Siddiqui and Sukhramani 200, Maiden 2001).

It is important to note, however, that there is a slight difference between what is popularly known as industrial social work and the discussion in this chapter. Social work practice in the water sector is a bit different. While social workers may administer welfare services to the employees in a water ministry, more importantly they work with local communities in awareness creation, planning and implementation of water projects. Through a bottom-up and participatory approach, social workers are also pivotal in preparing communities to engage in sustainable projects. Unfortunately, social workers in Kenya have been playing such roles not within the ministry of water but with NGOs that often collaborate with the ministry.

Social Work Training and 'Technical Fields'

As noted above, most of the core staff in the state's ministry of water are 'technical' staff trained either at universities or at middle-level colleges such as the Kenya Water Institute (KEWI). This institute in particular is recognised in the Water Act of 2002 as the state's official institution responsible for training technicians for the water sector.

By and large, social work training in Kenya does not prepare graduates to work in industrial establishments or in areas that would be considered 'technical' fields such as the water sector. While sociology students are exposed to such courses as 'sociology of work and industry', social work students are mainly taught and exposed to service delivery at social work-oriented departments and organisations. To start with, social work training is mainly generic, with students being equipped with skills in the five social work methods, that is, casework, group work, community organisation, social welfare administration and research.

Casework (counselling) prepares students to work in hospitals (where counselling for those infected with HIV/AIDS is a major component), mental institutions or with street families or in family therapy. The group work method equips students with skills tailored to be functional in many rural areas or slums where self-help groups are a prerequisite to assistance from development organisations. Channelling assistance through groups is seen as a more viable way of reaching many needy people in a single intervention and with higher chances of sustainability. Suffice it to say that some of these groups are 'water groups' with the sole aim of accessing safer water particularly in the rural areas.

In a poor country such as Kenya, community organisation/development skills are necessary to enable a social worker know how to mobilise communities for larger development programmes and empower the people for self-determination. Social welfare administration is also a critical component in the training of social workers in Kenya. Students are exposed to the challenges of managing welfare as well as development programmes and the need to blend 'field' skills with 'office' skills. The basic message is that while social programmes need social workers to serve individuals, groups and communities directly, these programmes also require administrators as well.

The general trend in Kenya, however, is that administrative positions in social work organisations are hardly held by social work graduates. Many of these organisations tend to have a bias towards job-seekers with skills in management, accounting or economics. This trend has forced some social work students to pursue courses in management or accounting alongside their social work course or upon completion of social work training. Finally, social work students are expected to acquire research skills. At the University of Nairobi, for example, social work students are not only taught quantitative and qualitative research methods but also social statistics as well. The students are made aware that in practice, social workers need to conduct research on the changing needs of client systems, analyse data, write reports and disseminate findings.

Field attachments while in college or university shape the kind of work social workers will do once they graduate. In Kenya, many social work students do their practicum in what could be referred to as the 'traditional' social work organisations such as the children's department, non-state child welfare agencies, children homes, self-help groups (mainly for women and youth), hospitals (medical social work) and mental health institutions. In a country where there are about 8,000 NGOs, social work graduates see these organisations as their potential employers. And with over 2.5 million[3] orphaned and vulnerable children (OVCs), there are very many NGOs that provide services for children and where many social workers work. These range from multilateral ones such as UNICEF, international NGOs like World Vision, Plan International, Child Fund and Save the Children to local NGOs like Mama Ngina Children's Home in Nairobi. Besides, most social work graduates (particularly degree holders) expect to work in towns either to run programmes in the slums or to coordinate community development activities in the rural areas from the comfort of the city. This fixation with the city makes it difficult for social workers to explore working opportunities in rural areas where a number of organisations are supporting communities to access clean water. But one cannot entirely blame the students for preferring jobs in towns. The curriculum in training institutions is partly to blame among other factors such as

3 PLAN (2010), citing UNICEF estimates.

accommodation, ethnicity and transport challenges in most rural areas, leave alone the hard-to-reach regions.

In addition to the foregoing, very few social work students see commercial establishments such as banks, manufacturing firms or government ministries (e.g. water) as places where field practicum could be carried out and as potential employers. And although most of these organisations have welfare departments, 'industrial social work' does not apply here as many organisations do not see social workers as potential employees with the right skills. Most human resource and welfare departments in Kenya's corporate sector are run by graduates of business schools possessing either degrees in management or human resources.

Social Workers Uneasy Role

As noted so far, the role of social workers in facilitating access to clean water is not clear cut. This can be attributed to the fact that social workers are not hired to work in the government's water ministry or water training institutes. Even in social work training institutions, social work students are rarely inducted into the field of water and sanitation. And this is surprising bearing in mind that these are common social work related challenges facing millions of people in Kenya. But as noted earlier, the water sector, perhaps because of the way the government ministry in charge of this docket projects it, is largely seen as 'technical' and hardly 'social'. While all and sundry will acknowledge that the ultimate users of water are individuals, groups and communities, social/community development workers play minor roles, if any. Although the Ministry of Water and Irrigation's (now Environment, Water and Natural Resources) mandate in the rural areas particularly is to provide water to communities, its core staff is virtually technical (e.g. hydrologists and civil engineers).

The ministry of water in a developing country where millions of people still have no access to clean water and where interventions are through communities that have contrasting cultures and values should have among its ranks social development workers. This would ensure that the needy communities are professionally identified, their problems well diagnosed and prioritised and the appropriate recommendations made. Besides, it is also the communities that eventually shoulder the burden of operation and maintenance of water projects once the ministry or 'donor' support comes to an end. And this is a responsibility that communities need to be prepared for by working with personnel possessing relevant competences in social development. But as far as the ministry's policy makers are concerned, social workers are expected to work with community organisations that the ministry collaborates with at the local level. They also stress that theirs is 'a very technical ministry' as one senior official put it when I asked him whether recruitment of social workers in the ministry is something the management would consider.

While he acknowledged that social workers 'could add value' in shaping of policies within the government and in social service delivery, he felt that this would 'water down' the ministry's 'core mandate'. Probing further whether he could see a role social workers could play at all, he quipped ' ... well, in the human resource section or welfare section ... I think they could do something there ... ' But I could see where he was coming from. Apart from the subordinate staff and those holding management positions, most of the ministry's core staff has always been drawn from civil engineers, hydrologists or water technologists. And while the ministry acknowledges the fact that the NGOs they collaborate with at the local level employ social workers, they do not see how a social worker would 'fit' in the ministry's hierarchy.

Worse still, many social workers do not perceive themselves as having a role in water supply. Most of them see their roles as anchored in the 'traditional' rehabilitative approach where social protection is limited to case work, child care, family welfare and social rehabilitation. And even where social workers possess developmental skills, few organisations, including government, recognise such competence. In government departments, social workers are deployed as medical social workers in hospitals (e.g. at Kenyatta National Hospital), children's officers and social development officers (Ministry of Gender, Children and Social Development), Probation officers (Ministry of Interior and National Coordination), social welfare officers (Kenya Prisons Service) or simply as 'social workers' (Ministry of Devolution and Planning). In the non-governmental sector, social workers have many roles including child development officer, project/programme officer, program assistant, project manager, community development officer, community development facilitator, community animator, social worker (e.g. in a children's home), monitoring and evaluation officers, food security specialist, etc.

It is important to note here that even in the NGO sector, the role of social workers in the provision of water is not clear. Some organisations employ social workers only after recommendations in a monitoring and evaluation (M&E) report and usually when the project has faced relevance, efficiency and sustainability problems. In many water and sanitation projects, the critical aspects of community participation, need diagnosis, need prioritisation, attitudinal change, capacity building and sustainability are ignored. Partly because of this, Kenya is littered with stalled or collapsed water projects. To mobilise communities, many development organisations rely on local administrators rather than social workers. The other problem is duplication of services where organisations offering the same services (e.g. water supply) are over-concentrated in some areas while other equally needy ones are neglected. This is partly because many organisations do not see the need for carrying out baseline surveys before they embark on implementing their programmes.

On a positive note though, some local organizations and bilateral ones (e.g. SIDA and DANIDA), or international NGOs such as World Vision, recognise the need to integrate 'hardware' (technical) and 'software' (social dynamics) in water provision. In these organisations, the technical aspects of water supply are blended with social development work and environmental conservation.

Challenges and the Way Forward

As alluded to above, there are a number of explanations as to why social workers are not playing their rightful role in the water sector. This has partly to with social work's tradition and the way it is perceived. Just like in many others parts of the world, social work in Kenya is perceived mainly as a rehabilitative intervention and not developmental. This perception influences the way social work students are trained and the kind of jobs they look for once they graduate. There is need therefore not only to revitalise social work training but also to engage the relevant institutions where social workers work or could work in this process. Blending theory and practice would significantly change the current state of affairs.

Another critical aspect has to do with the non-recognition of social work as a profession. Both government and NGOs do not recognise social workers as professionals. In government for instance, the starting salary scale for a social work degree holder is job group 'J' while 'professionals' such as economists, accountants and engineers start as job group 'K'. Besides, jobs that should be performed by social workers are given to graduates in sociology, anthropology, political science or even literature. Since the government should be the pacesetter, NGOs have adopted a similar trend in remuneration and deployment. The non-recognition of social work as a profession perhaps explains why the ministry of water may not see what difference social workers would make if they were to join the ministry's ranks.

The non-recognition of social work as a profession is also compounded by lack of specialisation in training. As noted, much of the social work training in Kenya is generalist, not specialised. There are few social workers in Kenya with specialised training in areas that would be relevant in water and sanitation, e.g., project management or community development. This is compounded by the fact that there are many training institutions offering training that is wanting in quality. In some cases, those who teach may have no background in social work. As this calls for due attention, there is also need to embrace new developments in social work and social development in both training and in practice.

Some of the challenges identified above would be addressed by a professional association. While there is the Kenya National Association of Social Workers (KNASW), it is for all practical purposes, moribund if not dead. That means that social workers do not have a forum where they can meet and exchange ideas, the profession has no visibility and there is no regulation of what social do. One way this problem could be addressed is by coming up with an act of parliament on social work and community development. This would give social work the legal mandate that is lacking at the moment. It is in such an act that social work curricula, who social workers are and where they can work (including the water sector), would be stipulated.

A revitalised social work would assist the government to achieve her future plans in the water sector. For instance, Kenya aims to provide clean and potable water at source less than 1 km in high potential areas and less than 5 km in Arid and Semi-arid Areas (ASALs). By 2010 coverage was projected to be 85 per cent urban and 75 per cent rural (Kenya 2008). That has not been achieved. The ministry of water may need to embrace some of the ideals in medical hospitals. Although hospitals' core staff are doctors, clinical officers and nurses, social workers have always worked in these institutions to deal with other critical non-physical (e.g. psychosocial) aspects that patients face. This blending of the 'technical' and the 'social' is perhaps what should be adopted in a water ministry.

Overall, there is need for social workers and the government to do more. While social work voices are discernible where there are issues of gender inequality, domestic violence, food shortages, and child neglect, one hardly hears about social workers when it comes to discussions on access to clean water.

References

Day, D. (1996) "Water as a Social Good", Australasian Journal of Environmental Management, Vol. 3, 1, 26–41.

Kenya, Republic of. (2002) The Water Act 2002, Nairobi: Government Printer.

Kenya, Republic of. (2008) Kenya Vision 2030: A Globally Competitive and Prosperous Kenya, Nairobi: Government Printer.

Kenya, Republic of. (2010) Population and Housing Census 2009, Nairobi: Government Printer.

Kenya, Republic of. (2010) Kenya Constitution 2010, Nairobi: Government Printer.

Kenya Water for Health Organisation (KWAHO). (2009) Enhancing Water and Sanitation Governance in Kenya: Human Rights Based Approach to Reforms in the Kenya Water Sector, Nairobi: KWAHO.

Maiden, R.P. (2001) Global Perspectives of Occupational Social Work, New York: Haworth Press, Inc.

Nyanchaga E.N. and Ombongi, K.S. (2007) "History of Water Supply and Sanitation in Kenya 1895–2002" in Environmental History of Water, edited by P.S. Juuti, T.S. Katko and H.S. Vuorinen. London: IWA Publishing, 271–320.

PLAN. (2010) Breaking Barriers for Orphans and Vulnerable Children in Kenya, Nairobi: Kul Graphics Limited.

Siddiqui, H.Y. and Sukhramani N. (2001) "Occupational Social Work in India", in Global Perspectives of Occupational Social Work, edited by R.P. Maiden. New York: Haworth Press, Inc., 43–64.

UNICEF and WHO, (2012) Progress on Drinking Water and Sanitation: 2012 Update, USA: WHO/UNICEF.

Electronic sources

Kumar, D.K. (2006) Industrial Social Work: What to do with Organization? http://www.indianmba.com/Faculty_Column/FC311/fc311.html. 17/04/2013.

Water, Pesticides, and Our Children's Wellbeing: International Initiatives and Opportunities

K. Findley, L. Laiewski and M.E. Rogge

From tap water to the world's oceans, biological and chemical threats to the world's water are ubiquitous, of consequence and increasing (Alavanja 2009, Carr and Nealy 2008, Goldmann 2004, Kolpin, Thruman, and Linhart 1998, Ongley 1996, Palaniappan et al. 2012). This contribution documents the growing threat of contaminated water to humans and other species, with a focus on children's wellbeing and using data about one complex group of chemicals, pesticides. We critique and illustrate water source-focused initiatives and emerging principles designed to promote creative, sustainable solutions to this environmental problem. We draw on actions associated with local to international initiatives, introduce the Precautionary Principle, and incorporate other international data to present creative, feasible, interdisciplinary, international approaches that aim to reduce pesticide contamination in water sources and, therefore, to protect the world's children from the harmful effects of pesticide exposure.

The Nature and Scope of Pesticide Contamination in Water

The global output of pesticides and other chemicals is expected to be 85 per cent higher in 2020 than it was in 1995 (Hogue 2001). In many respects, the greatest potential for harm from pesticides is through contamination of water sources that support ecosystems, food chains, agriculture and other aspects of human community and other species, across rural to urban populations. Health risks from pesticides are magnified and concentrated through biological processes as these chemicals move up the food chain – as top of the proverbial food change, humans consume the most concentrated forms of these and other chemicals (Ongley 1996).

Globally, in 2007, consumers spent more than $39.4 billion to purchase about 5.2 billion pounds of pesticides (Grube, Donaldson, Kiely and Wu 2011). Where pesticides are used, people can be exposed through consuming contaminated foods and water, through exposure to skin or breathing in contaminated soil, air, and water (Goldmann 2004, Ongley 1996). Pesticides and their breakdown products can be transported long distances through the atmosphere to geopolitical regions where they are not used or produced (Hageman, Simonich, Campbell, Wilson, and Landers 2006; Muir, Teixeira, and Wania 2004, Palaniappan, Gleick, Allen, Choen, Christian-Smith, and Smith 2012). This ubiquitous distribution increases the likelihood of exposure of individuals in areas with little or no pesticide use.

Pesticides and many of their degraded breakdown components can be harmful to all species, including humans, in the natural environment (Carr and Nealy 2008,

Ongley 1996). Some pesticide breakdown products are equal to or more toxic than the parent substances. Some are very hard or impossible to remove; they persist in the environment for years and decades, contributing to increases in their frequency, concentration and potential to harm (Carr and Nealy 2008, Kolpin, Thruman, and Linhart 1998).

Consequences for Children

Exposure to pesticides and other chemicals has been documented to have harmful physical effects to children and adults (Robert and Karr 2012, Wigle et. al 2008). UNICEF (2008) estimated that organic and chemical water pollution kills two million children per year in developing nations.

Children are especially at risk from and sensitive to pesticide exposure. Developmental activity, including crawling and frequent hand-to-mouth activity, brings children into direct contact with surfaces where pesticides and other contaminants settle. Children are exposed to a higher concentration of pesticides than adults; pound for pound, children consume proportionately more food and water and breathe more air than adults (Schafer and Marquez 2012). As a child's internal organs – especially the brain – develop rapidly, cognitive and developmental processes can be altered significantly because of pesticide exposure.

Children also are exposed to pesticides through maternal contact with the chemicals, pre-natally and after birth. Pre-natal pesticide exposure is associated with structural changes in the human brain (Raugh et al. 2012). Ferriera et al. (2013) found maternal exposure to be associated with childhood leukaemia in Brazilian children, through altered genetic and molecular mechanisms. Mothers transmit vital nutrients to their children through breastfeeding; pesticides can block absorption of these nutrients and impair development processes. Primarily through food and water intake, certain pesticides, including certain organic chlorine compounds, concentrate more in human breast milk than in dairy cattle milk (Stefanidou, Maravelias and Spiliopoulou 2009). Pesticides in human breast milk have been documented worldwide (Pruss-Ustin, Bos, Gore and Bartram 2008).

Birth weight – influenced by genetics, the health of the mother during pregnancy, and environmental factors, including exposure to pesticides and other chemicals – is a predictor of childhood brain development and of a variety of health outcomes in childhood, adolescence and adulthood. Low birth weight is associated with neurological, neurosensory, and functional impairment (Vohr et. al 2000). In a study conducted in Denmark, children of mothers exposed to pesticides during pregnancy had significantly lower birth weight (Wohlfahrt-Veje et. al 2011). Wickerham et al. (2012) found associations between pesticides and stunted fetal growth among newborns in rural China in whom pesticides were detected in their umbilical cord blood.

Childhood exposure to pesticides has been linked to infant mortality and to a myriad of other physical and cognitive problems, including birth defects, developmental delays, childhood brain cancers, Autism Spectrum Disorders, Attention-Deficit/Hyperactivity Disorder and endocrine disruption (Schafer and Marquez 2012, Wigle et al. 2008). Thakur and colleagues (2009) examined the link between children's social functioning, language skills, premature birth and stillbirths in Punjab, India, a location with a high concentration of pesticides and other chemical contaminants.

In developed and developing countries, children in rural communities, whose families subsist through agricultural activities, are more susceptible to pesticide exposure (Robert and Karr 2012). Demographic and geographic factors worldwide do affect the degree, types, and sources of exposure through pesticide use in rural, suburban and urban uses for agricultural and pest management purposes. Pesticides have been shown to affect children through more subtle pathways for which rigorous research is just beginning to emerge. In westernized countries, for example, certain pesticides in household tap water have been associated with increases in food sensitivity and allergies in children (Jerschow et al. 2012).

The harmful effects of pesticide contamination in water have been documented internationally. New research continues to emerge that provides a better understanding and a larger base of evidence for understanding this problem and the short and longer term consequences for children, their families, and their communities.

Pesticide and Water Initiatives, Strategies and the Precautionary Principle

How we value children's health is represented, in part, by how we set priorities are set and make decisions. In the case of pesticide exposure, decisions are also part of how we frame environmental issues. The precautionary principle, rooted in German environmental policy, was developed as an international response to those concerns and adopted in the 1992 United Nations Conference on Environment and Development's Rio Declaration. (Kriebel et al. 2001). The Rio Declaration stated: 'Where there are threats of serious or irreversible damage, lack of full scientific certainty shall not be used as a reason for postponing cost-effective measures to prevent environmental degradation' (United Nations Conference on Environment and Development 1992). The four major components of the Precautionary Principle are (1) using decision-making in the face of uncertainty; (2) shifting burdens of proof, (3) fully analysing alternatives to potentially harmful activities, and (4) using democratic decision-making structures (Raffensperger and Tickner 1999). A fifth, widely applied component is that the consequences of a given action on the most vulnerable populations must take priority. This Principle has generated significant application in the biological, environmental and public health sciences, and substantial controversy worldwide in comparison to "traditional" risk assessment methods, such as cost-benefit analysis and quantitative risk assessment

Table 7.1 International Resources for Pesticides, Water and Children's Wellbeing

United Nations Sustainable Development Knowledge Platform	Central resource for UN Environmental Programme links and the ongoing Agenda 21 initiative. Action plans for member states with regard to multiple environmental issues, including pesticide use	http://sustainabledevelopment.un.org/
United Nations Food and Agriculture Organization (FOA)	Hosts the Joint Meeting on Pesticide Residues" (JMPR), an expert ad hoc body administered jointly by FAO and WHO. Data and other resources about agriculture and water pollution, pesticides and local to international alternatives	http://www.fao.org/agriculture/crops/core-themes/theme/pests/jmpr/en/
World Health Organization	Water Quality and Health Strategy (2013–2020) Healthy Environments for Children Alliance	http://www.who.int/water_sanitation_health/publications/2013/water_quality_strategy/en/index.html http://www.who.int/heca/en/
Organisation for Economic Co-operation and Development (OECD)	WWW Pesticide Sites in OECD Countries and Other Organizations	http://www.oecd.org/env/ehs/pesticides-biocides /www.pesticidesitesinoecdcountriesandotherorganisations.htm
International Agreements and Treaties on Pesticides	Extensive source on international treaties and conventions	http://www.epa.gov/oppfead1/international/agreements/
United Nations Decade on Education for Sustainable Development (2005–2015)	Local to international initiatives and strategies to educate the world's population toward environmental sustainability	http://www.unesco.org/new/en/education/themes/leading-the-international-agenda/education-for-sustainable-development/about-us/
International Decade for Action "Water for Life" (2005–2015)	Promotes international water and water-related improvements. Documents and recognizes outstanding best practices	https://www.un.org/waterforlifedecade/

United Nations Millennium Development Goals	Goals focus on most vulnerable populations and includes environmental sustainability. Local to international strategies by countries and development institutions.	http://www.un.org/millenniumgoals/
Pesticide Action Network International	IGO in over 90 countries and six regions working to replace pesticides with "ecologically sound and socially just alternatives."	www.pan-international.org/
CropLife International	IGO with extensive integrated pest management (IPM) information. IPM is used extensively in rural, suburban and urban areas to reduce pesticide use	http://www.croplife.org/ipm
United States Environmental Protection Agency	Federal government agency links to international agreements and protocols, local to international data, initiatives and projects	http://www.epa.gov/pesticides/health/index.htm
Children's Environmental Health Network	USA-based NGO addresses policy, research and preventive strategies	http://www.cehn.org/
Beyond Pesticides	USA-based NGO provides research, policy, and practice information on pesticides and alternatives to their use.	http://www.beyondpesticides.org/

(Grandjean 2004, Jostmann 2007, Phillips 2006, Sadeleer 2007). Social workers worldwide would gain from assessing its utility, particularly with regard to children and environmental threats to their wellbeing.

"Environmental" ("ecological", "green") social work is gaining broader understanding as an essential component of our person-in-environment perspective and core values (Besthorn 2012, Gray, Coates and Hetherington 2012, Hoff and McNutt 2004, Rogge 2008). In 2012, the *Global Agenda for Social Work and Social Development* (2012) signalled a worldwide embrace of the relevance of environmental threats to our constituents and profession, as the International Association of Schools of Social Work, the International Council on Social Welfare and the International Federation of Social Workers established environmental sustainability as one of the Agenda's four pillars for action.

The chart above contains resources about policy, practice and research on pesticides and water at local, national and international levels. The resources include international and national governmental resources; international (IGO) and national (NGO) nongovernmental organizations; the United Nations and World Health Organization; a resource for accessing international treaties and conventions with regard to pesticides; and three international initiatives. Resources specific to the threat to children's wellbeing from exposure to pesticides and other environmental contaminants are included. Through these links, resources, action plans and strategies can be identified for each region of our world, and for the majority of nations.

Through expanding our environmental knowledge, skill and resource base with regard to critical concerns such as the contamination of the world's water by pesticides and other chemicals, our social work profession can increase further our relevance and effectiveness worldwide. We can; therefore, more fully secure the wellbeing and human rights of children – and their families and communities – in a natural environment that is indeed sustainable.

References

Alavanja, M.C.R. (2009) Pesticide use and exposure extensive worldwide. *Reviews on Environmental Health,* 24(4), 303–309. Retrieved from: http://www.ncbi.nlm.nih.gov/pmc/articles/PMC2946087/.

Besthorn, F.H. (2012) Deep Ecology's contributions to social work: A ten-year retrospective. *International Journal of Social Welfare*, 21:3, 248–259.

Carr, G.M. and Neary, J.P. (2008) Water quality for ecosystem and human health. 2nd Edition. United Nations Environment Programme Global Environment Monitoring System. Retrieved from http://www.unep.org/gemswater/Portals/24154/publications/pdfs/water_quality_human_health.pdf.

Eskenazi, B., Rosas, L.G., Marks, A.R., Bradman, A., Harley, K., Holland, N., Johnson, C., Fenster, L., and Barr, D.B. (2008) Pesticide toxicity and the developing brain. *Basic & Clinical Pharmacology & Toxicology*, 102, 228–236.

Ferreira, J.D., Couto, A.C., Pombo-de-Oliveira, M.S., Koifman, S. (2013) In utero pesticide exposure and leukemia in Brazilian children < 2 years of age. *Environmental Health Perspectives*, 121(2), 269–275.

Gamble, D.N, and Weil, M. (2009) *Community Practice Skills: Local to Global Perspectives.* New York: Columbia University Press.

Global Agenda. (2012) *Global Agenda for Social Work and Social Development: Commitment to Action.* International Association of Schools of Social Work, the International Council on Social Welfare and the International Federation of Social Workers. Retrieved from http://www.globalsocialagenda.org/.

Goldmann, L. (2004) Childhood Pesticide Poisoning: Information for Advocacy and Action. Prepared for the United Nations Environment Programme. Retrieved from: Grandjean, P. (2004) Implications of the precautionary principle for primary prevention and research. *Annual Review of Public Health*, 25, 199–223.

Gray, M., Coates, J., and Hetherington, T. (Eds) (2012) *Environmental Social Work*. Routledge Press.

Grube, A., Donaldson, D., Kiely, T., and Wu, L. (2011) Pesticides industry sales and usage: 2006 and 2007 market estimates. U.S. Environmental Protection Agency; Washington (DC). Retrieved from: http://www.epa.gov/opp00001/pestsales/07pestsales/market_estimates2007.pdf.

Hageman, K.J., Simonich, S.L., Campbell, D.H., Wilson, G.R., and Landers, D.H. (2006) Atmospheric deposition of current-use and historic-use pesticides in snow at national parks in the western United States. *Environmental Science & Technology*, 40(10), 3174–3180.

Hoff, M.D., and McNutt, J.G. (Eds.) (2004) *The Global Environmental Crisis: Implications for Social Welfare and Social Work*. Aldershot: Avebury/Gower House.

Hogue, C. (2001) Responding to globalization. *Chemical & Engineering News, 79*(18), 11.

Jerschow, E., McGinn, A.P., de Vos, G., Vernon, N., Jariwala, S., Hudes, G., and Rosenstreich, D. (2012). Dichlorophenol-containing pesticides and allergies: Results from the US National Health and Nutrition Examination Survey 2005–2006. *Ann Allergy Asthma Immunol,* 109(6):420–5.

Jostmann, T. (2007) Precautionary principle for toxic chemicals – no alternative to safeguard societal benefits. *Human & Experimental Toxicology*, 26, 847–849.

Kolpin, D.W., Thruman, E.M., and Linhart, S.M. (1998) The environmental occurrence of herbicides: The importance of degradates in ground water. *Archives of Environmental Contamination and Toxicology*, 35, 385–390.

Kriebel, D., Tickner, J., Epstein, P., Lemons, J., Levins, R., Loechler, E.L., Quinn, M., Rudel, R., Schettler, T., and Stoto, M. (2001). The precautionary principle in environmental science. *Environmental Health Perspectives, 109*(9), 871–876.

Muir, D.C.G., Teixeira, C., and Wania, F. (2004) Empirical and modeling evidence of regional atmospheric transport of current-use pesticides. *Environmental Toxicology and Chemistry*, 23(10), 2421–2432.

Ongley, E.D. (1996) Control of Water Pollution from Agriculture-FAO Irrigation and Drainage Paper 55. Food and Agriculture Organization of the United Nations: Rome. Retrieved from: http://www.fao.org/docrep/W2598E/W2598E00.htm.

Palaniappan, M., Gleick, P.H. Allen, L., Cohen, M.J., Christian-Smith, J., and Smith, C. (2012) Water quality, in *The World's Water Volume 7: The Biennial Report on Freshwater Resources*, edited by P.H. Gleick. Springer.

Phillips, M.L. (2006) Registering skepticism: Does the EPA's pesticide review protect children? *Environmental Health Perspectives, 114*(10), A593-A595.

Pruss-Ustin, A., Bos, R., Gore, F., and Bartram, J. (2008). *Safer Water, Better Health: Costs, Benefits and Sustainability of Interventions to Protect and Promote Health*. ISBN 978 92 4 159643 5. Geneva: World Health Organization.

Raffensberger, C., and Tickner, J. (Eds) (1999) *Protecting Public Health and the Environment: Implementing the Precautionary Principle.* Washington, DC: Island Press.

Rauha, V.A., Pererab, F.P., Horton, N.K., Whyatt, R.M., Bansale, R., Haoe, X., Liue, J., Barrf, D.B., Slotking, T.A., and Petersone, B.S. (2012) Brain anomalies in children exposed prenatally to a common organophosphate pesticide. *Proceedings of the National Academy of Sciences, 109(20), 7871–7876.*

Robert, J.R., and Karr, C.J. (2012) Council on Environmental Health. Technical report: Pesticide exposure in children. *Pediatrics*, 130(6):e1765–e1788.

Rogge, M.E. (2008, 2012) Environmental Justice, in *Encyclopedia of Social Work 20th Edition*, T. Mizrahi and L. Davis (Editors-in-Chief). Oxford University Press. Volume II,. 136–139.

Sadeleer, N. (2007) Implementing the precautionary principle: Approaches from the Nordic Countries, EU and USA. *Reciel*, *17*(1), 137–139.

Schafer, K.S., and Marquez, E.C. (2012) *A Generation in Jeopardy: How Pesticides are Undermining our Children's Health & Intelligence.* Oakland, CA: Pesticide Action Network North America.

Stefanidou M, Maravelias C, and Spiliopoulou C. (2009) Human exposure to endocrine disruptors and breast milk. *Endocr Metab Immune Disord Drug Targets*, 9(3):269–76.

Thakur, J.S., Prinja, S., Singh, D., Rajwanshi, A., Prasad, R., Parwana, H.K., and Kumar, R. (2009) Adverse Reproductive and child health outcomes among people living near highly toxic waste water drains in Punjab, India. *J Epidemiol Community Health,* doi:10.1136/jech.2008.078568. Retrieved from *http://jech.bmj.com/content/early/2009/10/19/jech.2008.078568.*

United Nations Children's Fund (UNICEF) (2008) *UNICEF Handbook on Water Quality*. New York: (UNICEF).

United Nations Conference on Environment and Development. (1992) *Agenda 21: Programme of Action for Sustainable Development; Rio Declaration on Environment and Development; Stated Forest Principles: The Final Text of Agreement Negotiated by Governments at the United Nations Conference on Environment and Development*. June1992, Rio de Janeiro, Brazil.

Vohr, B.R., Wright, L.L., Dusick, A.M., Mele, L., Verter, J., and Jean J. (2000) Neurodevelopmental and functional outcomes of extremely low birth weight infants in the National Institute of Child Health and Human Development Neonatal Research Network, 1993–1994. *Pediatrics, 105:6 1216–1226.*

Wickerham, E.L., Lozoff, B., Shao J., Kaciroti N., Xia Y., and Meeker, J.D. (2012) Reduced birth weight in relation to pesticide mixtures detected in cord blood of full-term infants. *Environmental International*, 47(2012), 80–85.

Wigle, P.T., Arbuckle, T.E., Turner, M.C., Burube, A., Yang, Q., Liu, S., et al. (2008) Epidemiological evidence of relationships between reproductive and child health outcomes and environmental chemical contaminants. *Journal of Toxicological and Environmental Health, Part B: Critical Reviews*, 11(5–6), 373–51.7

Wohlfahrt-Veje, C., Main, K.C., Schmidt, I.M., Boas, M., Jensen, T.K., Grandjean, T., Skakkebæk, N.E. and Andersen, H.R. (2012) Lower birth weight and increased body fat at school age in children prenatally exposed to modern pesticides: A prospective study. *Environmental Health,* 2011, 10:79.

Chapter 8

Social Economy and Sustainable Social Development: Local Lessons – Global Challenges

Social economy has emerged as a grass-roots entrepreneurial, not-for-profit sector on the basis of people's needs and democratic values. Social economy has an impact on growth, employment creation and welfare endeavours to enhance social, economic, and environmental conditions of communities. Social economy comprises a kind of buffer between the market economy private enterprise and the public sectors. It also furthers new forms of partnerships between public, private and nonprofit actors. Institutional diversity may be as important as biological diversity for our long-term survival.

Microfinance System and Rural Housing: Generation of a Model from the Case Study of Sivagangai District, Tamilnadu, India

Francis Adaikalam

Overview of Rural Housing in India

A plethora of International Declarations, Special Reports on Housing by UN and Indian Constitutional schedules mandated the State to ensure shelter for its citizens to achieve wellbeing (UDHR 1948, Habitat Agenda 1996, WHO 2011). In India it is a challenge to turn this commitment into reality. The Working Group on Rural Housing, Planning Commission of India for the 12th Five Year Plan period (2012–2017), estimates 43.67 million housing shortage units in rural areas. These mostly, i.e. 39.30 million units (90%), pertain to the Below Poverty Line (BPL) population for a total population of 1.21 billion (Census, 2011). People who belong to the BPL category come from lower social group categories such as Scheduled Tribe – 16.4 per cent; Scheduled Caste – 28.5 per cent and Other Backward Communities – 40.1 per cent (National Sample Survey Organization, 2008). Tables 8.1 and 8.2 give an interesting picture of rural housing reality in India. Impressively, more than 90 per cent of the rural households live in an owned house compared to 69 per cent of urban households. But they live mostly in *katcha* (thatched) or semi-*katcha* (semi thatched) households. The challenge here is how

Table 8.1 Occupancy by Residence for India

Census	Rural			Urban			Total		
	Owned	Hired	Others	Owned	Hired	Others	Owned	Hired	Others
1991	94.6	4	1.5	63.1	34.1	2.8	86.3	11.8	1.8
2001	94.4	3.6	2.1	66.8	28.5	4.7	86.7	10.5	2.8
2011*	94.7	3.4	1.9	69.2	27.5	3.3	86.6	11.1	2.4
NSS rounds									
50 (1993)	29 ucca	38.1semi-Pucca	38.40 utcha						
51 (1994–1995)	93.4	3.9	2.7						
56 (2000)	38.6 ucca	31.6 semi-Pucca	30 utcha						
60 (2004)	94.5	3.4	2.1						
62 (2005–2006)	93.4	3.8	2.7	60.0	34.9	5.1			
63 (2006–2007)	94.4	3.4	2.2	62.8	33.8	3.8			
64 (2007–2008)	94.4	3.3	2.3	61.5	32.2	5.3			
65 (2008–2009)	95.0	3.3	1.6	61.5	33.2	5.3			
	55.4ucca	27.6 semi-Pucca	17.0 Kutcha						

Source: Census 2011; *NHB (2012); & Rao, K. Hanumantha. ed. (2012). & Various NSS rounds

one creates structures and policy intervention to address the 43.67 million housing shortage in India. Adequate housing needs must be understood as instrumental for poverty reduction and realising social justice.

Housing Finance in India

National Housing Bank (NHB), which regulates Housing Finance records; 'Affordable Housing has serious implications for the country's housing sector as the wide gap between the demand and availability of housing units increasing' (NHB, 2012). The Indian Planning Commission in its 11th Five year period document (2007–2012) emphasised financial inclusion as the prime objective to fill the gap. NHB 2012 reports that the formal credit system is concentrated in semi-urban and urban areas since the outstanding amount by the banks in these areas is 16 per cent and 28 per cent respectively compared to 7.5 per cent in rural areas. NHB (2012) attributes the reasons for this trend to lack of proper titles, lack of collateral, largely unorganized sector, low income households and fear of becoming a non-performing asset. Hence, other alternative models of financing were to be explored in India such as Housing Microfinance which so for focussed on livelihood, social function and other emergency needs (Srinivasan. N. 2010 page 19).

Study Objectives and Methodology

This contribution is written from a study funded by National Housing Bank (NHB), a subsidiary of Reserve Bank of India, in 2008. It is primarily to understand the field reality of housing micro-finance, its dynamics and suitable policy interventions in the Sivagangai District of Tamil Nadu.

The primary data was gathered between May and August 2008. It used qualitative method in an exploratory design in collecting data. Focused group discussions, in-depth interviews and a semi-structured questionnaire were used to elucidate their response. The stakeholders interviewed using snowball sampling were: Government representatives (18 Nos); financial institution members (10); Non-Government Organisations (7); and individual beneficiaries (92) totalling 127 individuals. Secondary data were collected from various sources such as government departments, banks and the Internet in the period between June 2008–December 2008 and May 2012.

Case of Sivagangai District: Generation of Model

Sivagangai District is an interesting case for generating housing micro-finance for various reasons such as the type of housing, agrarian workforce (77%), industrially-backward districts and BPL population (Potential Linked Credit Plan,

Table 8.2 Distribution of Households by Type of Census Houses Occupied, 2001

Housing Category	Sivagangai District	Chennai District	Villupuram	Tamil Nadu	All India
Total Households (Nos.)	280,607	827,811	632,884	14,173,626	191,963,935
Permanent (%)	57.61	87.16	37.34	58.52	51.80
Semi-Permanent (%)	26.54	7.18	9.25	18.15	30.04
Temporary (%)	15.85	5.61	53.39	23.31	18.14
Unclassifiable	00.00	00.05	00.02	00.02	00.02

Source: Census 2001

X Five year Plan Period 2002–2007, NABARD). Administratively, the District is divided into six *Taluks*, 12 Blocks and 431 *Panchayat* Villages. The type of household living in a permanent structure as shown in Table 8.2 is almost the same as that State-wide. Also the number of people living in semi-permanent and temporary structures is not as alarming as in the case of Villupuram District where it stands at almost 63 per cent.

Housing Requirement in Sivagangai District

The District Rural Development Agency (DRDA) data on government housing schemes show that there was an almost 50 per cent drop (2,394 in 2004 to 1,427 in 2007) in providing housing for marginalised communities from 2004 to 2007 (Annual Statistical Handbooks 2004–2007, Sivagangai District). In the district, those living in semi-permanent and temporary housing constitute 26.54 per cent and 15.85 percent of the population respectively, which in real numbers is 1,14,000 (rounded off). Amongst these, according to DRDA, more than 90,000 households belong to the BPL category. The total number of households in the State of Tamilnadu is 1,41,73,626 (Census, 2001) and the housing shortage is 1.98 million (rural 0.44 million and urban 1.54 million).[1] If we extrapolate the State figure into the District, which has 2,35,304 households the housing shortage stands as 18,000 for the reference year 2001 (Adaikalam, 2009). The district's housing shortage in the year 2008 was around 39,000 which includes a minimum housing estimation (11,000) and replacement demand (28,000). In the year 2011 it is projected as 54,000 (MHE 24,000 + RD 30,000) (Adaikalam, 2009).

1 Taken from NBO report. These estimates have been arrived at by utilizing the data released by RGI on the basis of Census of India, 2001.

Analysis of Various Schemes for Housing in Sivagangai District

Indira Awas Yojana (IAY)

Housing for the rural poor, popularly known as *Indira Awas Yojan a* (IAY), has its own strengths and struggles in the district. Money allocated to each district is disbursed to DRDA by the Government and it follows a roster system to choose beneficiaries. A sanctioning letter is sent to the beneficiaries once the list is finalised by DRDA, based on the BPL census. The beneficiary can obtain a loan amounting to INR 20,000 from a nearby bank by producing a copy of a) a land deed in his/her name b) a Public Distribution Card and c) a letter of communication from DRDA (Adaikalam, 2009). All documents and the sanctioning letter from the commercial bank must be submitted to the Block Development Office (BDO) which is the local implementing and monitoring authority. The bank loan is given in two instalments as indicated in Figure 8.1. A joint liability group (JLG) formed by the borrowers will guarantee the bank's return and the interest rate for this group is fixed at 4 per cent for seven years.

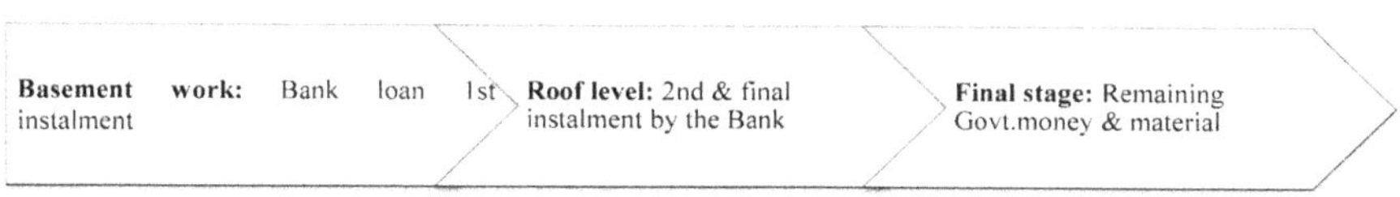

Figure 8.1 Process of Money Imbursement for IAY scheme

Source: Field notes

In Sivagangai District, 60 per cent of the IAY beneficiaries belong to a Scheduled Caste population and 40 per cent to Other Caste Groups (Adaikalam 2009). Though the number above looks quite impressive and has helped people to have house collie workers and elderly were notably left out. One of the major reasons for this misplaced allotment is reliance on the poor figures from the BPL survey in 2002 (field notes). Loans sanctioned by the banks helped people to complete the construction of houses compared to previous years but they have also become a deterrent for many families as the cost of building is doubled. Better supporting mechanisms among the Backward Classes helped them to complete construction compared to the Scheduled Caste. Transparency and sharing of information about the scheme would reduce wastage of time, energy and money (daily labour) for the beneficiaries.

Dynamics of other Housing Models in Sivagangai District

Case 1. Agency X, operating from a town called Tirupattur started a rural housing project as part of the Food for Community Development – a subsidy combined

Table 8.3 Various housing models developed by an Agency X [Field visit notes. 2008]

	Reference period	**Total cost of the house**	**Contribution 1**	**Contribution 2**	**Contribution 3**
TYPE 1 Food For Community Development (FFCD)	1982–1990	Rs. 1 lakh	Rs. 75,000 from VA	–	Rs. 25,000 Beneficiary
TYPE 2 Credit Cum Subsidy Scheme (CCSS)	2003	Rs. 40,000	Rs. 10,000 Subsidy by DRDA	Rs.30,000 Bank loan	–
TYPE 3 Kutcha House Improvement Programme	2004	Rs.10, 000	Rs. 2,500 Subsidy by DRDA	Rs.7, 000 Bank loan	Rs. 500

Source: Reproduced with minor changes Adaikalam (2009)

with partnership from people in the form of manual labour, finance, materials and planning made the scheme successful.

Case 2. Voluntary Agency Y started based on Gandhian–Sarvodaya ideology in Tamil Nadu. Initially it began as a cooperative vision for landless people but later focused on integrated rural development. They implemented three types of housing schemes. 1. **Group Housing:** Funds from housing finance companies are refinanced for group housing. 2. **Housing Promotion Fund:** Roofing, plastering mud walls and flooring, electrification of semi-*pucca* houses, construction of toilets and bathroom facilities. The fund is routed through agency patronage self-help groups. 3. **Credit-Cum-Subsidy Programme:** to build new homes with a unit cost per house of Rs.40,000. It is implemented through a tripartite partnership model with beneficiaries (Rs.10,000–20,000) + Bank (Rs.20,000) + VO (Rs.20, 000).

Conclusion and Recommendation

This contribution tries to capture various models in housing micro-finance. Table 8.4 provides good understanding of the success of some models over others. In-depth impact assessment should be done on the outcomes of various models operated in the district. One of the hurdles to getting loans in rural areas is the

Table 8.4 Developing a Model by comparison schemes implemented by Government and Voluntary Agencies

S.No	Particular	IAY	Voluntary Agency X	Voluntary Agency Y
1	Work Execution	Contractor Driven	Beneficiary Participation	NGO & Beneficiary Participation
2	Credit	Government Subsidy & Bank loan	Community Assesses individual borrower need. Subsidy & Loan is provided accordingly	Fewer Subsidies. Banks are linked
3	Type of House	Very little floor space	Spacious	Less Spacious
4	Model of the house	Government specification & Govt. Engineer plans	Individuals' own choice	VA proposed model
5	Cost	Determined by market rate and contractor	Cost effective since beneficiary provides Manual Labour & locally avaliable raw materials	Cost depends on the market
6	Quality of House	Quality cannot be ensured since contract driven	Excellent quality ensued	To an extent if beneficiary fully involved themselves
7	Volume of Houses	Thousands	Hundreds	Hundreds
8	Social Consciousness	Allotment is made based on a nuclear family	Existing Family System is considered based on beneficiaries preference	Joint families are divided into nuclear families
9	Bank's obligation	Provides the minimum loan of INR.20,000/- at DRI rates to meet the targets.	VA tie up with Bank and banks are advised to suit beneficiary needs	Less or no linkage. VA acts as a creditor and takes the repayment risk.
10	Beneficiary contribution	Less contribution except margin money	Contribution in terms of supply of raw materials and human resource	Beneficiaries' contribute financially. Repayment long years

need to prove title to land which needs a serious policy change. RRBs and housing co-operative societies should be tapped in addition to the SHG-Bank scheme run by NABARD for housing micro-finance. Institutions can set aside a 'Grievance Day' for customers/prospective customers exclusively to address grievances about housing such as the Old Age Grievance Day. Such an arrangement will not only smooth the loan process but also pave the way to mitigating legal hurdles. In order to encourage beneficiaries who make regular repayments a special incentive scheme needs to be introduced wherein a certain percentage of the interest paid can be provided as cash-back.

References

Adaikalam, F. (2009) District-wise Plan for Housing in the Districts of Sivagangai and Pudukottai, Tamilnadu. http://www.nhb.org.in/R&D/FINAL.pdf.

Srinivasan, N. (2010) Microfinance India. New Delhi: Sage Publications. http://www.microfinanceindia.org/uploaded_files/publication/1311572030.pdf Internet materials accessed on 31 May 2013.

UN (2006): Report of the UN Special Rapporteur on adequate housing as a component of the right to an adequate standard of living, Miloon Kothari, E/CN.4/2006/41, 21 March 2006.

WHO (2011): Housing WHO Library Cataloguing-in-Publication Data. Housing: shared interests in health and development .(Social determinants of health sectoral briefing series, ISBN 978–92–4-150229–0. http://whqlibdoc.who.int/publications/2011/9789241502290_eng.pdf. http://www.censusindia.gov.in.

NHB (2012) Trends and Progress report on Housing http://nhb.org.in/Publications/Report-Trend-and-Progress-of-Housing-in-India-2012.pdf).

NHB (2008) Draft Report Housing Microfinance, National Housing Bank. July 2008.

Kumar, R. (2005) Banking with the Poor: NABARD's Experience, http://www.microfinancegateway.org/gm/document-1.9.27276/22948_Paper_on_SHG_Bank_Linkage_pakistan_conference.pdf.

Rao, K.H. ed. (2012) Rural development statistics 2011–12, 2012 (August). National Institute of Rural Development, Ministry of Rural Development, Government of India. http://www.nird.org.in/Rural%20Development%20Statistics%202011–12/data/rdscontents.pdf.

NSSO round reports 50 (1993), 51 (1994–1995), 56(2000), 60 (2004), 62 (2005–2006), 63 (2006–2007), 64 (2007–2008), 65 (2008–2009).

NEAER Report (2010) Roof Above the Head: A Qualitative Assessment of Rural Housing in India: http://www.eaber.org/sites/default/files/documents/NCAER_Bhide_2010.pdf.

NABARD on rural Housing: Web resources: (25 May 2013)

http://www.censusindia.gov.in/2011census/hlo/Slum_table/Slum_table.html

http://mospi.nic.in/mospi_new/upload/statistical_year_book_2012/htm/Ch-28%20Housing/Housing-writeup.pdf
http://mospi.nic.in/Mospi_New/upload/SYB2013/ch28.html
http://mospi.nic.in/Mospi_New/site/home.aspx
http://mospi.nic.in/Mospi_New/upload/MDG_Report_2011_12july12.pdf
http://www.nabard.org/nonfarm_sector/pdf/chapter%202.pdf
http://www.thehindu.com/features/homes-and-gardens/home-finance/rural-housing-scheme/article4393381.ece
http://www.thehindu.com/todays-paper/tp-features/tp-propertyplus/rural-housing-scheme/article4395416.ece
http://www.indiahousingstat.com/housing/17/housingschemes/121/stats.aspx
http://knowledge.wharton.upenn.edu/india/article.cfm?articleid=4219
http://www.constructionweekonline.in/article-4767-housing_rural_india/2/
http://www.constructionweekonline.in/article-4767-housing_rural_india/
http://news.oneindia.in/2008/11/25/indias-rural-housing-stock-to-rise-up-to-44-mln-by-2025-report-1227628977.html
IAY scheme website- GoI: http://iay.nic.in/netiay/home.aspx
Rural Development Statistics- NIRD: http://www.nird.org.in/Rural%20Development%20Statistics%202011–12/rdsindex.html
http://rural.nic.in/sites/downloads/programmes-schemes/DRDA_workshop_details.pdf
http://www.performance.gov.in/sites/default/files/document/strategy/RD.pdf
http://www.igidr.ac.in/FSRR/PDF/20130202EPW_regulatingmicrofinanceinstitutions.pdf
CM Solar Powered Green House Scheme TN: http://www.tnrd.gov.in/schemes/st_cmspghs.html
Housing in Tamilnadu: http://censusindia.gov.in/Data_Products/Library/Housing_TN_link/housing_tamilnadu.pdf

Enterprising Social Work: Social Enterprise Action Research with Remote Indigenous Communities in Australia

Deirdre Tedmanson

Introduction

Social enterprises are market-based activities which provide social benefits, through engagement of people in productive activities that meet community needs or foster profits for redistribution as social investment for other community development ventures. While the role of social enterprise in aiding social functioning is well-established in the enterprise literature (Dees and Anderson 2002, Nyssens 2006, Nicholls 2008), its importance as a community development platform for fostering emotional and social wellbeing is under-discussed in social work. Factors such as historical exclusion from 'mainstream' markets; absence of a sustainable local economy; lack of skills training; institutionalised racism; poor governance; and tensions between social, cultural and economic goals (Altman 2001, Dana and Anderson 2007) are often cited as challenges which impede Indigenous communities wishing to develop livelihoods on their own lands, on their own terms. However, strengths-based approaches can facilitate community participation in strategies to overcome the disempowering effects of the 'welfare economy' (Pearson 2000). Social entrepreneurship builds social capital which supports health and wellbeing (Helliwell and Putnam 2004) and shifts focus from individual pathologies towards social relationships and community processes in producing social and emotional wellbeing (Sonn and Fisher 1998, Hunter 2004, Anderson 2007, Ingamells et al. 2011). This contribution draws on empirical insights emerging from long-term participatory action research being conducted in partnership with elders, communities and homelands in the Anangu Pitjantjatjara Yankunytjatjara Lands of central Australia, which is exploring social enterprise development in remote community contexts. It argues that participation in social enterprise development has a range of benefits: including strengthening family networks, enhancing trust and promoting self-reliance, social esteem and cultural safety (Tedmanson and Guerin 2011, Taylor and Guerin 2010). It suggests social entrepreneurship enhances interdependence and independence and provides a platform for meaningful social work engagement through community development. It suggests also that over time mental health and social and emotional wellbeing dividends accrue to communities engaged in self-determined social entrepreneurial activities.

Participatory Action Research

Particpatory Action Research or PAR as it is known, is an emancipatory method of inquiry which has its early conceptual roots in Freire's liberation pedagogy (1970) and social psychology theory (Lewin 1948). In essence, those affected by an issue or problem work on activities of practical relevance to their lived experience to generate new understandings of both process and context. It is a method that enables the *co*-production of new knowledge and theoretical insights for innovation and social change in applied contexts (Tedmanson and Banerjee, 2010). It is a form of community development-based research which aims to make explicit the power inequalities between professional researchers and those researched by fostering processes of collaboration so that value is added in and learning by doing is facilitated through cycles of reflexivity.

In Indigenous contexts relations of power and the continuing effects of past colonial violence and racial domination, raise ethical and epistemological concerns for social workers engaged in community development and community-based research. Indigenous research theorist Linda Tuhiwai Smith (1999; see also Denzin et al. 2008) conceives of participatory action methods as empowering ways to decolonise research engagements. Accountability for real world impacts of research on those affected is central to this method. Cycles of reflection–action–reflection form what Freire calls 'conscientisation', in which consciousness and consciences are developed through processes determined by Indigenous peoples operating as partners and not being positioned as objects or subjected to external gaze (Tedmanson and Banerjee 2010). Knowledge is *co*-produced and rooted in solidarity with the everyday experiences and issues that shape the political and social realities of Indigenous life.

In this project PAR was adopted as an appropriate approach in direct response to the expressed wish of Indigenous A<u>n</u>angu Pitjantjatjara Yankunytjatjara (APY) communities, located in remote central Australia, who enlisted researchers to work *with* them on ways to foster sustainable local social enterprises on *their* lands. The project was thus a collaborative partnership between three A<u>n</u>angu community partners, comprised of homelands and extended kinship groups; a mainstream community development NGO active in the region and a coalition of researchers from social work, social science and business disciplines, working together to help generate community based social enterprises. A<u>n</u>angu community members identified the sorts of social enterprises they wanted to develop in their local area and exchanged ideas, expertise and cultural knowledge with researchers about priorities and possibilities. Through collaborative workshops, interviews and group discussions, direct actions were undertaken to identify and trial enterprises ranging from cultural tourism ventures led by local elders, to homeland-based sole operator soap-making, the potential for a local community-based paper products venture was explored and a second-hand clothes store was plotted in different locations. Learning in this process comes from doing and research understanding emerges through a reflexive analysis of constraints and opportunities encountered in undertaking each endeavour. Simplistic externally-produced 'white' notions, which often construct Indigenous community lifestyles as dysfunctional and unproductive, start to rapidly unravel when people work together towards common goals. Learning becomes a two-way process and the ignorance and prejudice that shapes Indigenous domains is revealed in the struggles, isolation, lack of basic services and hardships encountered. By 'walking' together, new ways of 'being' together are found and notions of top-down capacity building give way to the richness of capacity sharing (Tedmanson, 2012). Through this enterprise development project a new partnership has also been struck through which community members operate an academic cultural tourism venture entitled 'Learning Country' which aims to enable social work academics and students to participate in an 'on country', pre-placement learning exchange experience, hosted and guided by elders and leaders from the Anangu homelands. The community is paid for their services as social work educators engaged in inducting and orientating future practitioners, in the Anangu remote community context. Understanding grows, 'both ways' knowledge builds and the social enterprise, as well as all involved, gains benefit.

Empowering Social Work

Empowerment potentially provides a useful framework for more reflexive social work practice with Indigenous Australians (Whiteside et al. 2011, Fredericks 2009). The 'ability to integrate Aboriginal knowledge alongside Western Paradigms' (Bennett et al., 2011: 30) remains a key to good practice in this context, and the development of ongoing respectful relationships are essential if empowerment gains are to be sustainable. The ways that external social, economic and health pressures can disempower communities is best revealed through deep and sustained processes of engagement, in ways that are active and kinesthetic as well as cognitive. In this social enterprise development project, periods of action were augmented by shared discussions to align people's experiences, observations and perceptions which helped to build shared understandings (Tedmanson 2012, Harms et al. 2011). Strengths-based approaches to community development emphasise the development of social capital and its leverage as a means for bridging to external resources (Poulin 2009). Strengths-based approaches reinforce local talents and build local capacity, fostering resilience and establishing a firm platform for local resistance to outside pressures that erode cultural efficacy. By focusing on local priorities and culture as asset and strength (rather than deficit) pride, confidence and motivation are enhanced. Support for and facilitation of, locally determined processes stimulates greater participation and lessens dependence on external approaches to 'solving' local community issues (Tedmanson and Guerin 2011). As national Indigenous leader Calma (2011:147) points out: 'by shifting the focus to human rights, social workers can move towards more empowering, rights-based practice that develops individual and community capacity'.

Enterprising Social Work

When people are enabled to work together for an achievable and sustainable common purpose which aids their independence or communal interdependence, social and emotional well-being is enhanced (Taylor 2006, Dockery 2009). Social enterprise development can bring benefits at multiple levels. It provides opportunities for small-scale ventures to emerge which can help to substitute locally grown or sourced produce for imported products. In the case of small scale cultural tourism it can augment homeland incomes and provide ways for local people to present cultural stories and ways of being as an asset to be proud of and which 'outside' people will pay to visit and engage with. Rather than relying on 'passive welfare' the communities engaging in this social enterprise project demonstrated competence and confidence in their active engagement in learning by doing, working into new markets and encouraging others to step into leadership and embrace opportunity. Participation in local social enterprise development has many interrelated nuanced layers of small but powerful benefits which if well planned, enacted collaboratively and instigated from a strengths based perspective, can have a lasting impact.

References

Altman, J. (2001) *Sustainable Development Options on Aboriginal Land: the Hybrid Economy in the 21st Century*. Discussion paper no. 226. Canberra: Centre for Aboriginal Economic Policy Research, Australian National University.

Anderson, I. (2007) Policy processes, in *Social determinants of indigenous health, edited by* B. Carson, T. Dunbar, R.D. Chenhall and R. Bailie. NSW: Allen & Unwin, 231–46.

Banerjee, B. and Tedmanson, D. (2010) Grass Burning Under Our Feet: Indigenous Enterprise Development in a Political Economy of Whiteness. *Management Learning* 41(2): 147–65.

Bennett, B., Zubrzycki, J. and Bacon, V. (2011) What Do We Know? The Experiences of Social Workers Working Alongside Aboriginal People, *Australian Social Work*, 64:1, 20–37.

Borzaga, C. and Defourny J. (2004) *The Emergence of Social Enterprise*. London: Routledge.

Calma, T and Priday, E. (2011) Putting Indigenous Human Rights into Social Work Practice, *Australian Social Work*, 64:2, 147–155.

Dana, L. and Anderson, R. (2007) *International Handbook of Research on Indigenous Entrepreneurship*. Cheltenham: Edward Elgar.

Dees, J. and Anderson, B. (2002) For-profit Social Ventures. *International Journal of Entrepreneurship Education* 2(1): 8–21.

Denzin, N.K., Lincoln, Y.S. and Smith L.T. (2008) *Handbook of Critical and Indigenous Methodologies,* Los Angeles: Sage Publications.

Dockery, M. (2009) 'Culture and wellbeing: The case of Indigenous Australians', Centre for Labour Market Research Discussion Paper Series (09/01). Perth: Curtin Business School, Curtin University of Technology.

Fredericks, B. (2009) Which Way that Empowerment? Aboriginal women's narratives of empowerment. *AlterNative: International Journal of Indigenous Scholarship*, 4, 6–19.

Freire, P. (1970) *Pedagogy of the Oppressed*. New York: Herder & Herder.

Harms, L., Middleton, J., Whyte, J., Anderson, I., Clarke, A., Sloan, J., Hagel, M. and Smith, M. (2011) Social Work with Aboriginal Clients: Perspectives on Educational Preparation and Practice, *Australian Social Work*, 64:2, 156–168.

Helliwell J. and Putnam R. (2004) The social context of wellbeing, *Philosophical Transactions B*, 2004 September 29; 359(1449): 1435–1446.

Hunter E. (2004) Commonality, difference and confusion: changing constructions of Indigenous mental health. *Australian E-Journal for the Advancement of Mental Health* (AeJAMH) 446–7984 Vol. 3 no. 3 Dec 2004.

Ingamells, A. Holcombe, S and Buultjens, J. (2011) Economic development and remote desert settlements, *Community Development Journal*, vol. 46, no. 4, 436–457.

Lewin, K. (1948) *Resolving social conflicts; selected papers on group dynamics.* New York: Harper & Row.

Nicholls, A. ed. (2008) *Social Entrepreneurship: New Models of Sustainable Social Change*. Oxford: Paperback Edition, Oxford University Press.

Nyssens, M. ed. (2006) *Social enterprise – at the crossroads of market, public policies and civil society.* London: Routledge.

Pearson, N. (2000) *Our Right to Take Responsibility.* Cairns: Noel Pearson.

Poulin, J. (2009) *Strengths-Based Generalist Practice: A Collaborative Approach,* 3rd edition London: Cengage Learning.

Smith, L.T. (1999) *Decolonizing Methodologies: Research and Indigenous Peoples,* Dunedin, NZ: University of Otago Press.

Sonn, C. and Fisher, A. (1998) Sense of community: community resilient responses to oppression and change. *Journal of Community Psychology,* 26: 457–472.

Taylor, J. (2006) '*Indigenous peoples and indicators of well-being: an Australian perspective on UNPFII global frameworks*', Working Paper No. 33/2006, Canberra: Centre for Aboriginal Economic Policy Research, Australian National University.

Taylor, K and Guerin, P. (2010) *Health Care and Indigenous Australians: Cultural Safety in Practice*. South Yarra: Palgrave-MacMillan.

Tedmanson, D and Banerjee SB. (2010) Participatory Action Research, in *Encyclopedia of case study research*, edited by Mills, A.J., Durepos, G., and Wiebe, E. Thousand Oaks, CA: Sage Publications.

Tedmanson, D. and Guerin, P. (2011) Enterprising social wellbeing: social entrepreneurial and strengths based approaches to mental health and wellbeing in "remote" Indigenous community contexts *Australasian Psychiatry*, 19 S 1, 3–33.

Tedmanson, D. (2012) Whose Capacity Needs Building?, in *Against the Grain, Advances in Postcolonial Organizational Studies*, edited by Prasad, A. Copenhagen: Copenhagen Business School Press, 245–275.

Whiteside, M, Tsey, K and Earles, W. (2011) Locating Empowerment in the Context of Indigenous Australia. *Australian Social Work*, vol. 64, no. 1, 113–129.

Chapter 9

Sustainable Social Development: Research, Education and Practice

Social Development was launched at the Copenhagen UN summit, 1995, with *Sustainable Development* being the follow-up in Johannesburg, 2002. Resource conservation, economic development and social justice were seen as the most important aspects of the concept of sustainability. The integration of the three pillars of Economic development, Environmental development and Social protection seem to be the urgent challenge now. Where do we stand in research? What kind of implementation is reached in the curriculums of social work? We also need examples of social work practice.

Social Work Global Access Model: Promoting Social Equity and Social Change

Monit Cheung and Rhonda G. Patrick

Web-facilitated education provides task-oriented ways to efficiently maximise learners' understanding of social issues (Du, Havard, and Li 2005). When done globally, it is typically initiated by a university to connect students from another country through web-based technology (Crowell and McCarragher 2007). The medical and scientific scholars have already created information access for international scholars (Burton 2011). However, students in countries with limited technology infrastructure have restricted access to scientific resources and limited connections with likeminded colleagues and educators across continents, from rural and emerging countries (Burton 2011; Darkwa and Mazibuko 2000; Hürlimann et al. 2011).

Since global justice issues affect direct and policy practice, social workers must learn from various cultural perspectives to enhance the globalisation functions of their practice (Regan and Youn 2008, Zondiros 2008). This contribution focuses on the design of a global access education model using technology to connect both urban and rural universities from six continents. Its purpose is to identify the strengths and obstacles related to providing educational opportunities in social work through technology and other channels.

The Global Access Education Model

The "open university" concept has become increasingly popular given its promise to accommodate non-traditional student learning and online teaching (Al-Fadhli 2011). However, many online courses do not provide a face-to-face intercultural environment that connects students globally (Edwards 2011). In a global educational environment, several components and requirements must be added to this "open" concept.

A first requirement is linguistic competency that goes beyond language because learning can be transmitted through technologically enhanced practice dialogues (clinical or macro) that focus on both verbal and nonverbal connections and for both individuals and communities. Selwyn (2011:85) states that technology has become the "grammar" in higher education. The reality is that social work educators must address "linguistic diversity" in global learning (Askeland and Payne 2007:161). In this project, a teaching channel is called "linguistic appreciation" which encourages students to present practice dialogues in their native tongue as part of cultural sensitivity training. Students must achieve a basic level of translinguistic understanding through interpreting the content (through intake information and body language) and the process of the presented dialogues from the other culture.

Second, the concept of "shepherd leadership" illustrated by Chiu (2009:42) can be used to identify peer learning modalities that are culturally appropriate. Each virtual learning community must take up the role of a "shepherd" to create cultural learning tools for practice designs and discussions on developing culturally effective interventions.

Third, curriculum contents must address local needs and contextualised practice on topics caused by discrimination and prejudices in various cultural and ethnic communities (Fischer 2007, Ziegahn 2005). These concepts have been discussed by Rotabi, Gammonley, Gamble, and Weil (2007) to include empowerment, interdependence, reciprocity, social change, and transnational and cultural competence. Students can exchange additional topics relevant to the local needs.

Finally, the course structure must be collaborative and flexible. The global access framework must include "flexible peers" (pairing with peers from selected countries), "structured topics" (developing structured learning objectives for goal attainment measures), and "cognitive collaborations" (moving from "adoptive" to "adaptive" learning and teaching) (see Du et al. 2005, 207).

Delivery Method

Twelve universities have been invited to participate in this project. They represent countries from six continents: Africa, Australia, Asia, Europe and Oceania, South and North America. Target issues include those that have been influenced by social change and affected global economy, human capital, hunger, disaster management, and social work practice. This global web-based education model yields innovative ideas for comparing urban and rural practice, and emphasises the importance of virtual learning processing, community participation, and exchange of service implementation strategies that are culturally relevant.

Step 1: Presentation of Global Access Ideas

The authors presented the ideas related to setting up a global classroom at the World Social Work Conference in Sweden. Faculty from China (including Hong Kong), Denmark, Israel, Japan, Mexico, Poland, Russia, South Africa, and the United States, from both urban and rural locations, will work together in this collaborative project of creating global social justice strategies through "virtual classroom" and "global learning". In this process, a global access classroom will be set up to address issues affecting social work practice in the world.

Step 2: Website Development

Figure 9.1 Global Access Website
Source: http://rgpgcsw.com website developed by Patrick and Cheung (2013); Photos in the header are provided by the website theme package

With a specific website design (www.rgpgcsw.com), we have provided a global link for participating universities, titled "Educational Partners", to learn from each other and to have accessibility to university websites to learn about highlighted social issues and course participants from each country. With this method, we are not limited to the initial group but can extend our invitation to the globe to include additional partners. Students and faculty facilitators may post video blogs, podcasts, and other media related to social justice in respective countries. This real-time sharing method will help social workers understand the needs of humankind and be motivated towards examining global social justice issues.

Step 3: Blackboard Learn Development

The authors have worked with the University of Houston (UH) information and technology experts (Tammy Hoskings and Georgetta Michko) to set up a course shell through the university's cyber environment. Students and instructors from other parts of the world can apply a POI (person-of-interest) identification and enrol in Blackboard Learn without charge, provided that the first author will be their faculty sponsor. Once enrolled, the students and faculty from our educational partners can access course materials and interact through Blackboard discussion boards. In addition, a public access website will be available so that students can instantaneously exchange ideas. This component requires internet access but discussion materials can be downloaded to facilitate learning. Each university designs a virtual classroom that involves a video production of locally-based social work practice and direct interviews of students, instructors, and agency/community representatives with a focus on cultural perspectives on service implementation.

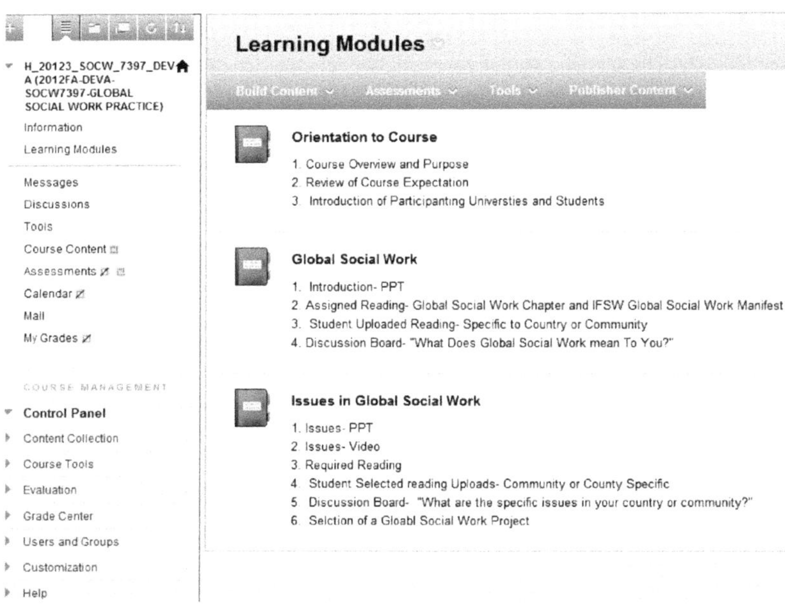

Figure 9.2 Blackboard Learn: Shell Development

Source: University of Houston Blackboard Learn course shell developed by Cheung (2013); use with permission

Step 4: Global Access Materials

Teaching materials, video blogs and interviews of community practitioners will be included in the discussion to identify: 1) definitions of social, political and economic justice; 2) global access to social work practice learning; 3) practice principles that are contextualised, indigenous, value- and culture-based; 4) political and economic challenges; and 5) micro and macro interventions in the global environment. Evaluations must be conducted throughout the experience. Moving beyond course goals and objectives, instructors must participate in an integrative discussion at the end of each semester to incorporate student inputs for future course planning and development. Globalisation is used as a learning linkage, not only as course content.

Next Steps

Although past experiences have revealed limitations regarding internet use (Rovai and Downey 2010), our team has used other technological arrangements (such as phone calls initiated from the host country) to ensure students' access to creative material transfers such as practice dialogues. In this design, a global classroom is established in which the students from the six continents bring multidimensional cultural perspectives into critical-thinking discussions about equity issues (Chiu 2009). This global access model includes various components such as locale assessment (urban or rural), creativity in presenting social-system variations (contextualised practice modalities), technological and practice tutoring (for both students and instructors), marketing of the course, and university endorsement. Limitations are to be assessed: physical boundary, financial constraints, recruitment challenges, quality assurance, faculty development in virtual classroom setup and implementation, course design and pedagogy. Through mutual support particularly between students with and without technological support, we can provide technological consultation opportunities and also address disproportionality issues affecting indigenous populations.

Implications to Sustainable Social Development

Kusakabe (2012) supported the idea utilising global access to create bonding, bridging and bracing social capital to achieve sustainable development. In the development process of this global justice course, we not only recognised the essentialness of the three steps suggested by Kusakabe but also acknowledged the need for a fourth step through this course development experience: 1) bonding – establishing a relationship by finding and joining with other partner universities around the world to sustain a mutual support environment; 2) bridging – connecting students, faculty, practitioners and community stakeholders to share global issues and then prioritise community needs through this data sharing process; 3) bracing social capita – creating opportunities for students to develop social work skills and to improve interpersonal communications through global interactions; and 4) balancing – enhancing community empowerment and local development through actual world experiences while simultaneously promoting principles of contextual practice.

These four steps, in their own ways, provide educators a means to support research on the effectiveness of testing through various teaching methods with diverse populations in the world. The results will attest that this global access teaching model can be incorporated in all three areas of social work education – teaching, research, and practice. Szabó (2011:253) suggests that the concept of sustainable development should be communicated such that 'the continuous improvement of quality of life' becomes 'the concern of the entire world'. Therefore the creation of global classrooms is just the beginning on strengthening the concept of sustainable social development. The expansion and success of this global access project requires the recognition of its purpose by students, educators and practitioners around the world. With this, the global dialogues continue.

References

Askeland, G.A., and Payne, M.M. (2007) Distance education and international social work education. *European Journal of Social Work*, *10*(2), 161–174.

Al-Fadhli, S. (2011) Factors influencing the acceptance of distance-learning: A case study of Arab Open University in Kuwait. *International Journal of Instructional Media*, *38*(2), 147–157.

Burton, A. (2011) Sharing science: Enabling global access to the scientific literature. *Environmental Health Perspectives*, *119*(12), A520–523.

Cheung, M. (2013) *Global social work practice (online course)*. Retrieved from University of Houston Blackboard Learn: http://www.uh.edu/blackboard/.

Chiu, Y. (2009) Facilitating Asian students' critical thinking in online discussions. *British Journal of Educational Technology*, *40*(1), 42–57.

Crowell, L.F., and McCarragher, T.T. (2007) Delivering a social work MSW program through distance education: An innovative collaboration between two universities, USA. *Social Work Education: The International Journal*, *26*(4), 376–388.

Darkwa, O., and Mazibuko, F. (2000) Creating social work virtual learning communities in Africa. *Professional Development*, *3*(1), 15–24.

Du, J., Havard, B., and Li, H. (2005) Dynamic online discussion: Task-oriented interaction for deep learning. *Educational Media International*, *42*(3), 207–218.

Edwards, B.L. (2011) Social work education and global issues: Implications for social work practice. *Education*, *131*(3), 580–586.

Fischer, K. (2007) To connect with foreign students, Champlain College taps into technology. *Chronicle of Higher Education*, *54*(10), 44.

Hürlimann, E., Schur, N., Boutsika, K., Stensgaard, A., de Himpsl, M.L., Ziegelbauer, K., and Vounatsou, P. (2011) Toward an open-access global database for mapping, control, and surveillance of neglected tropical diseases. *PLoS Neglected Tropical Diseases*, *5*(12), 1–11. doi:10.1371/journal.pntd.0001404.

Kusakabe, E. (2012) Social capital networks for achieving sustainable development. *Local Environment, 17*(10), 1043–1062. doi:10.1080/13549839.2012.714756

Patrick, R., and Cheung, M. (2013) Global social justice. Website developed by authors at http://rgpgcsw.com.

Regan, J.C., and Youn, E.J. (2008) Past, present, and future trends in teaching clinical skills through Web-based learning environments. *Journal of Social Work Education*, *44*(2), 95–115.

Rotabi, K.S., Gammonley, D., Gamble, D.N., and Weil, M.O. (2007) Integrating globalization into the social work curriculum. *Journal of Sociology and Social Welfare*, *34*(2), 165–185.

Rovai, A.P., and Downey, J.R. (2010) Why some distance education programs fail while others succeed in a global environment. *Internet & Higher Education*, *13*(3), 141–147.

Selwyn, N. (2011) Digitally distanced learning: A study of international distance learners' (non)use of technology. *Distance Education*, *32*(1), 85–99.

Szabó, Z.K. (2011) Analysis of research on sustainable development the goals of sustainable development, practical and theoretical framework in EU and Romania. *Juridical Current, 14*(4), 253–261.

Ziegahn, L. (2005) Critical reflection on cultural difference in computer conference. *Adult Education Quarterly*, *56*(1), 39–64.

Zondiros, D. (2008) Online, distance education and globalisation: Its impact on educational access, inequality and exclusion. *European Journal of Open, Distance and E-Learning, 2008*(1), webpages. Retrieved from http://www.eurodl.org/index.php?p=archives&year=2008&halfyear=1&article=302.

Chapter 10

Environmental Justice at the Heart of Social Work Practice: Greening the Profession

Lena Dominelli[1]

Introduction

Environmental injustice, or the failure to share the earth's resources equitably, is rooted in environmental degradation that is caused by the normal processes of industrialisation and disasters, whether natural or (hu)man made. Unsustainable models of development, unequal power dynamics and unequal distribution of resources, central to the current global socio-economic system of neo-liberalism, exacerbate structural inequalities and affect poor and low income people most. They also undermine people's capacity to mitigate risk and deter the enhancement of resilience amongst low income people and their communities. Environmental justice relies on the equitable sharing of both benefits and the burdens involved in maintaining the healthy and sustainable environments that all living things can enjoy. Although everyone has a role to play in maintaining the earth's environment for the use of current and future generations of people, flora and fauna, social workers have particular roles to play in assisting the mobilisation of people and resources when they wish to declare that environmental degradation is unacceptable and aim to formulate alternative models of socio-economic and environmental development rooted in the principles of environmental justice and a healthy, sustainable existence.

This chapter focuses on social workers' roles in promoting environmental justice through a form of practice that I have called *green social work*. In making my case, I explore how disasters, including those generated by extreme weather events associated with climate change, aggravate vulnerabilities and undermine poor people's capacity to build resilience. I also expose the contribution to environmental degradation made by a socio-economic system, namely the current global form of capitalist neoliberalism that disregards the wellbeing of people, living creatures and the physical environment. And, I highlight the injustices and lack of sustainability that this model of growth perpetrates by highlighting the shortcomings of neoliberalism.

---

1 This Katherine A Kendall Award Lecture has been formatted to an accepted and published *Invited article* in the International Journal of Social Welfare 2013:22: 431–439.

Environmental Degradation: A Disaster that both Complicates and is Complicated by other Disasters

Environmental degradation is proceeding apace as ocean acidity rises, sea levels become higher as glacier ice melts, soil erosion proceeds at an unprecedented pace and the exhaustion of the earth's dwindling fertility reduces agricultural yields. Even setting aside land for agricultural purposes has failed to secure its future, and food insecurity is striding across the globe. Rising sea levels are increasingly likely to jeopardise the survival of small island states and threaten coastal cities, towns and villages. Coastal cities that are vulnerable to flooding have constantly growing populations as they absorb migrants from rural hinterlands and rising numbers of births, thereby making already vulnerable ecosystems even more fragile, for example coastal Bangladesh which has had many initiatives aimed at controlling coastal erosion through the planting of mangrove plantations. Klinenberg (2002) suggested how countries prone to heat waves can reduce deaths, especially among older people, by learning from the 1995 heat wave in Chicago.

Ill-health accompanies the destruction of healthy environments, especially as asthma and other respiratory diseases are triggered by air pollution inside and outside homes. For example, indoor pollution caused by the use of cardboard and dung has had 1.4 to 2 million deaths attributed to it annually. Additionally, outdoor air pollution kills 1 million people a year. Both are unacceptable from an environmental justice perspective because these features are found primarily amongst the world's most vulnerable populations in the Global South. Transfers of clean energy technologies could eliminate much of these deaths, and social workers could advocate at all levels from the local to the global for this to occur. In the UK, 30,000 older people die yearly of preventable respiratory illness due to cold, damp houses. Better housing for all, regardless of income levels, could become part of public policy that is then implemented at community level to reduce such unnecessary loss of life. As people's health is adversely affected by pollutants in the atmosphere, the water, the soil and in their houses, a holistic understanding of the issues and the possible repercussions is essential to reducing risk and asserting that environmental justice be included in planning housing, constructing commercial enterprises, building transportation, communication and sanitation infrastructures and in water supplies, and engaging in creating sustainable jobs. Environmental justice requires practitioners to tackle structural and individual forms of oppression that impact upon people and destroy the environment in the process of creating a privileged life for the few. The current unequal distribution of global wealth has produced 1,153 billionaires (i.e., people who have been recorded as having one or more billion dollars in 2010) while 3 billion people live on less than $2 per day (Kroll and Fass 2011).

Moreover, the costs of environmental degradation are high. Environmental degradation is itself a disaster that causes people to lose their homes and livelihoods, while the habitats that shelter plants and animals are destroyed. This state of affairs is not limited to the Global South. Western countries such

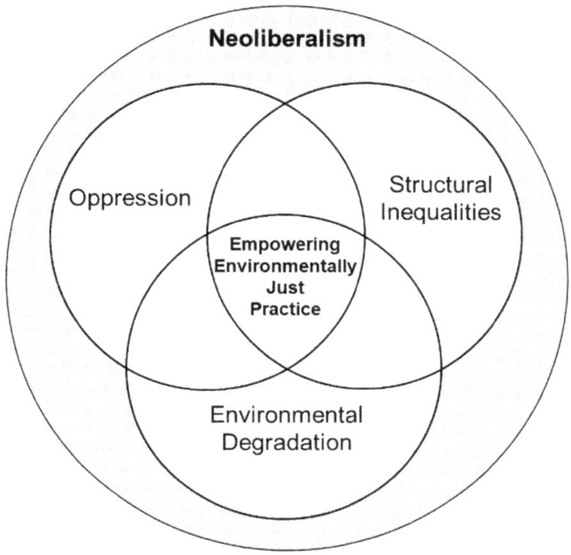

Figure 10.1 Empowering environmentally just practice

as Canada, the United States and Australia regularly experience losses linked to environmental degradation, particularly those linked to the exploitation of primary commodities. Some disasters that destroy the environment's capacity to renew itself are caused by people, and give rise to painful consequences for those affected. Others are attributed to 'Mother' Nature. Considered a 'natural' disaster, the 2004 Indian Ocean Tsunami killed 300,000 people. As in Hurricane Katrina in southern United States, children and older people died in disproportionate numbers. Assessing vulnerability and making the necessary preparations to reduce risk for such eventualities is necessary to eliminate these deaths (Schmidtlein, Deutsch, Piegorsch, and Cutter 2008, Wisner, Blaikie, Cannon and Davis 2004) and increase the resilience of people to respond to environmental catastrophes and reduce 'an ecology of fear' (Davis 1998). Failing to do so is an environmental injustice that reduces people's sense of having a right to be looked after by others as well as care for them (Laugier and Paperman 2006).

Types of disasters

Definitions of disasters are contested and there are many different kinds of disasters that impact on the physical environment and living things: 'Natural' disasters are defined as those that lack human intervention and include earthquakes, volcanic eruptions, tsunamis, landslides and floods. In contrast, (hu)man-made disasters include: poverty; armed conflict; industrial pollution; over-urbanisation in the form of mega-cities lacking adequate public health infrastructures and utilities; industrial accidents such as the one that occurred in Bhopal in India or Three

Mile Island in the USA; environmental crises caused by deforestation of tropical and temperate rainforests; and climate change. While a neat distinction between natural and (hu)man-made disasters is evident in the literature (see Evans and Manion 2002), the lines between these types of disasters are becoming increasingly blurred. Poverty, for example, aggravates the impact of all types of disasters, but the disaster literature seldom considers it a disaster in its own right (Dominelli 2012a). Furthermore, some scientists claim that climate change will lead to an increase in volcanic activity and earthquakes, especially if the Arctic and Antarctic ice sheets melt (Pearce 2012). Climate change also has the potential to make existing natural disasters worse, as super-storm Sandy demonstrated when it hit the USA in November 2012. Additionally, regardless of the genesis of a disaster, its impact will be evident in the lives of living things and the physical environment. The degraded environment that disasters leave behind will intensify the experience of environmental injustice felt by poor people.

'Natural' disasters can cause severe environmental degradation and initiate '(hu)man-made' disasters, for example the earthquake that produced the tsunami that damaged the nuclear reactors at Fukushima Daiichi in Japan in March 2011. The utilities company, Tokyo Electric Power Company (TEPCO), had been advised to either refrain from building that particular type of reactor in that spot or to strengthen its earthquake proof-ness. But this advice had been ignored for years under the guise of commercial interests linked to profitability, leaving the plant vulnerable (Lovins 2011). After the disaster, the Japanese government was compelled to step into the breach and provide resources when TEPCO found that it was unable to cover the costs and declared itself bankrupt (Sobie 2011). In this case, as in others when other nuclear reactors have broken down, for example Chalk River in Canada, Three Mile Island in the USA and Chernobyl in the Ukraine, human error has exacerbated mechanical malfunctions and increased the damage caused to people and the environment, often irreparably.

Additionally, the solutions utilised in the case of the breakdown of nuclear reactors, can store up other problems for the future, for example the pouring of kilometres of concrete to contain radioactive materials uses substances that deteriorate over time, thereby allowing radioactive materials and other heavy metals to seep into the water-table, causing further environmental damage and harm to living things (Zeller 2012). Limited redress in such cases increases people's frustrations about the lack of environmental justice for them as victims. For example, many survivors of Bhopal are still struggling for reparations.

The damage caused by disasters is embedded in everyday life practices or routines. Thus, they become part of the taken-for-granted fabric of peoples' daily existence and are not questioned either in terms of their impact or in the formulation of measures that can mitigate risk. In such situations, social workers have an important role to play in raising consciousness of such issues and mobilising people to do something about reducing vulnerabilities, enhancing resilience and securing environmental justice before a disaster takes place and afterwards. Preventative planning and action is essential in mitigating such risks.

Differentiated experiences of disasters

However, environmental degradation and disasters do not impact upon people equally. The differentiated outcomes of environmental degradation and disasters affect poor and marginalised people, poor regions and poor environments most. Bullard (1990) coined the word 'environmental racism' to describe the actions of industrial firms that dumped toxic substances on poor African-American neighbourhoods without their consent and destroyed their environments and health in the process. Moreover, countries in the Global South bear a disproportionate burden emanating from disasters, particularly so-called 'natural' disasters because: 76 per cent of the damages caused by disasters occur there; 92 per cent of people affected by disasters live in these countries; and 65 per cent of the economic losses caused by disasters are incurred there (IFRC, International Federation of the Red Cross and Red Crescent Societies 2009).

The damage caused to these localities is exacerbated by the poor state of the built infrastructure systems – transportation networks, communication links, power and utilities supplies and buildings including houses, hospitals, medical clinics and schools. The lack of availability and/or resilience in these structures is compounded by: the absence of important resources such as social networks and money for their development; the presence of corruption in governance systems that permit the misuse of existing resources; and the failure of insurance companies to provide affordable cover in poor communities. Additionally, these factors make it difficult for communities in the Global South to recover easily from the impact of disasters that have occurred. Furthermore, these problems make prevention and preparedness even more important for these locales and the groups of people living in them.

Other (hu)man-made disasters that damage the environment, injure people and destroy homes which are rarely considered as perpetuating environmental injustices involve aggravated violence, especially that perpetrated during armed conflicts. These contribute to environmental degradation through the ordinance that is fired and contribute to global warming as a result. Armed conflicts can also inhibit suffering people from receiving humanitarian aid or prevent them from reaching humanitarian aid camps, as occurred, for example, in the, Sudan (Toole 2006). Disputes in such situations may linger for years. The international community seems to be incapable of intervening to establish peace because the principle of national sovereignty over-rides the state's duty to protect its citizens/ residents (Dominelli 2012a). Syria's current civil war denotes an instance in which the international community has failed to bring a halt to hostilities while aid workers risk their lives to provide limited support for the people who are forced to leave their homes to save their lives. Once civilians have been evacuated, the destruction of homes and the physical environment proceeds rapidly. Another seemingly intractable illustration is the UN's failure to resolve the long-standing dispute over land and territory between Israel and Palestine. The periodic armed conflict that flares up between these two nations has seriously degraded the daily life and environment of Palestinians living in the 'Occupied Territories'. Social workers there are struggling both to establish the profession in extremely onerous circumstances and provide much needed services for victim-survivors, many of whom are children (Lavalette and Jones 2011).

Climate change itself, an important disaster, presages a type that blurs the boundaries between natural and (hu)man-made disasters because the cause of contemporary climate change is deemed to be human activity linked to the usage of fossil fuels (Dessler and Parsons 2006), but its impact on the environment is likely to occur through the storms and other 'natural events' that follow in its wake. Moreover, scientists predict growth in the frequency and intensity of extreme weather events including storms, heat waves, cold spells and the associated flooding and drought that produce higher levels of death especially amongst women, children and older people (Oven et al. 2011, Robine, et al. 2007). Poor people suffer most when the built infrastructure fails in heat waves and cold spells, too (Dominelli 2012b). The impact of climate change also impacts badly on the potential for the survival and recovery of poor people, for example African Americans in Hurricane Katrina (Pyles 2007). Although climate change is considered by many as the most crucial contemporary social issue that requires attention, there are many 'sceptics' (Giddens 2009).

Some disasters have silver linings. For example, the spread of some diseases can be controlled through fires, some of which occur naturally through lighting strikes that encourage regrowth in forests afterwards, for example the eucalyptus trees in Australia; or huge swathes of evergreen forests devastated by pests like the pine beetle in British Columbia, Canada. However, many fires each year are caused by people's behaviour – untended campfires, dropping litter, cigarette butts and arson. Wildfires caused by human negligence and natural events such as lightning strikes have caused millions of hectares of forests to burn in both temperate and tropical climes. Those started by people are all avoidable, while those caused by lighting are not. However, bringing both under control as quickly as possible is necessary to protect living things and the physical environment. Such considerations are included in the development of risk reduction strategies that aim to protect the environment and reduce the impact of environmental degradation on people.

Although all of the earth's residents experience the effects of climate change, their contribution to the problem and their experiences of it are differentiated. Unequal consumption patterns mean that rich people have contributed much more to greenhouse gases, but poor people experience their consequences more heavily: Currently, these disparities are indicated in the following realities, and except for the first item, impact most on people with low incomes:

- 1 billion people over-consume
- 1 billion people are dying of hunger
- 1 billion people are malnourished
- 1 billion people have no clean drinking water
- 2.6 billion people have no sanitation
- 1.6 billion people have no access to electricity

- 15 million people die of diseases annually because drugs that could cure them are not cheap and easily available
- Food prices are rising rapidly and in 2008 led to riots around the world (*Newsweek*, 4–11 June 2012:49)

These conditions exacerbate the impact of climate change on both the environment and people, especially during extreme weather events such as flooding, heat-waves and cold snaps, and expose how poverty intensifies the damaging effects of environmental injustice.

Children are particularly vulnerable during disasters. Unfortunately, some adults, for example paedophiles, prey on them for sexual exploitation. They often define children as orphans when they are not, and pretend to offer them assistance and care when they are grooming them for exploitation as occurred in Sri Lanka during the 2004 Indian Ocean Tsunami and in Haiti after the 2010 Earthquake. Getting children back to the routines of school is useful in preventing children from being taken advantage of by predatory adults, and also enhances their recovery.

Extreme weather events also create stresses on health and social care provisions if the built infrastructure fails, for example when the roads become impassable due to snow and ice and the lack of vehicles that are equipped to navigate slippery and icy terrain. This situation calls for egalitarian partnerships between the public and private sectors and informal carers living in the community, and are essential in mitigating the effects of environmental injustice. However, the link between formal and informal carers can become exploitative primarily of the women who provide the caring that is needed because state agencies begin to include the availability of informal and unpaid carers in their calculations of how extreme weather events will be catered for (Dominelli 2012b). Older people become particularly vulnerable in such situations – they lose formal services because providers are unable to reach them, and they become dependent on the goodwill of family which may live outside the area and neighbours. As a service provider interviewed in the BIOPICCC (Built Infrastructures and Health and Social Care Provisions for Older People Under Conditions of Climate Change) study claimed:

> when a person is being assessed we'll be asking about whether it is a family member or whoever, about what support they are willing and able to actually provide to that person. There is no sort of judgement made that someone must provide something for their loved one, but we would ask them because we don't want to take that care away either. But it is important from that kind of area that this is something we think about – it is part of the contingencies. (Dominelli 2012b: 8)

Additionally, people's expectations about receiving health care at the point of need are being revised as higher costs provide the reason that politicians use to justify demands for a greater financial contribution from the end user or the patient. These changes are becoming increasingly common in the West as they are being incorporated into policy planning processes during an age of austerity, as stipulated by policymakers and service providers in the quote given above. Moreover, policies of passing the costs onto marginalised groups can become exclusionary because poor people cannot afford to purchase medical care or other services normally provided through the welfare state. For example, people who cannot afford to buy medicines privately in Greece are protesting the impact of public expenditure cuts.

A crucial difficulty about extreme weather events is that while their propensity to increase has been predicted, exactly when and where they will strike and who they will encompass is less predictable. This makes being aware of the consequences of preparing for them (or not) complicated. In such circumstances, some community residents indulge in what Dupuy (2005) called 'enlightened catastrophism' by focusing only on what must be done today. Preparedness, to them, makes sense largely if the predicted disaster is imminent, not one that might not occur for years or even decades into the future. Such responses emphasise the need for consciousness-raising and publicity that will alert people to the potential dangers, highlight what steps they need to take to mitigate the risks they might face, be prepared to take action and know who and where to go to for help if disaster strikes. Social workers can play a crucial role in exploring such scenarios with local people in their communities and contribute to the realisation of environmental justice in the process. As practitioners, they can also highlight how destroying the environment through climate change creates winners and losers according to different scenarios that identify risk and mitigation strategies. Social workers can also act as translators who explain mathematical models to people in ordinary, non-jargonistic language that they can understand.

Investing in people through educational provisions is an important dimension in preparing for disasters. It also indicates points at which ensuring environmental justice coincides with securing social justice. Having an aware, well-educated populace is integral to achieving this outcome. However, this reality is far from being the case. Globally, more girls and women than boys and men are illiterate. This has implications for the extent to which women can exercise agency and engage in scientific discussions, especially if information is not made available to them via other means, for example through audio-visual means or photographs. Despite attempts to increase primary school education for girls and women as a key element in the Millennium Development Goals (MDGs) that aim to redress the literacy divide between and within countries, much remains to be done in this regard. Social workers have been involved in developing and promoting schemes that enable women to become educated, but need to do more to promote environmental justice in situations where women carry the major portion of the burden of caring for their families in distressed physical environments. This point is vividly portrayed by a woman villager interviewed by Dominelli (2012c:7) following the Indian Ocean Tsunami in Sri Lanka. She commented:

> If men's economic activities were hampered [by the Tsunami], they are excused by society … . Women are not excused by the society. They are supposed to

> do whatever … deal with the problems. They … take care of the household and the children.

Meeting development needs, particularly those linked to raising people out of poverty, can cause extensive environmental degradation if the development is not planned sustainably, and if poor people are not involved in multi-stakeholder partnerships to find solutions alongside experts. Sustainable development according to the Brundtland Report (1986) refers to the 'capacity to meet the needs of the present without destroying the potential of meeting the needs of future generations'. Regrettably, not all development follows sustainable principles. The issue of sustainability in the alleviation of poverty and in tackling climate change has become a highly disputed issue in the United Nations Framework for the Convention on Climate Change (UNFCCC) [Third world Network, TWN 2012]. Although agreement in this forum has not been forthcoming, no one and nothing will escape the impact of climate change. Unfortunately, the impact is variable, and those most vulnerable are those who have contributed the least to this phenomenon, for example small island states in the Global South that are in danger of sinking as sea levels rise (TWN 2012).

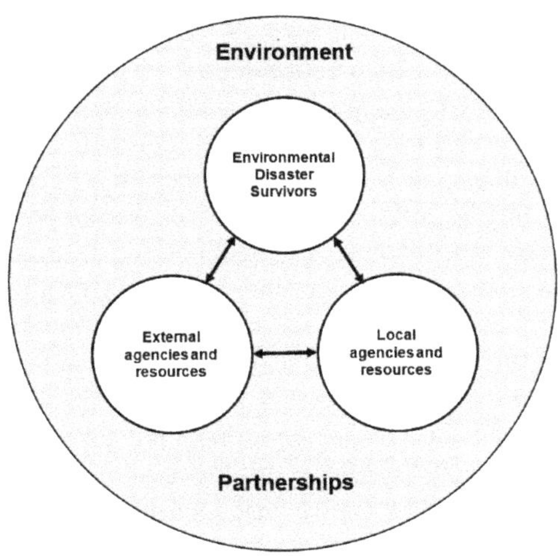

Figure 10.2 Multi-stakeholder environmental partnerships

Additionally, multinational firms are not held accountable by national governments or local people. Consequently, development can occur at the expense of the environment, causing hardship to various communities and damaging the physical environment. Indigenous people have been particularly hard hit by unsustainable mining practices, particularly those involving the extraction of gold, diamonds and other precious gems and valuable minerals such as tungsten. Mineral extraction endeavours have also led to armed conflicts in parts of Africa such as the Democratic Republic of the Congo (*War Child* 2012). Resolving these conflicts often involves multi-stakeholder partnerships composed of internal and external actors working alongside each other.

Indigenous people are often caught in the cross-fires between development that is often unsustainable and against the long-term interests of indigenous people committed to preserving their environment for the future. Mining for coal, gems and minerals is unsafe for people and environmentally destructive work, especially when driven by the profit motive. For example, in Latin America, the expropriation of lands, especially those of indigenous peoples, for development purposes, often done without their having a say in what happens or receiving compensation for what is lost, leaves them in precarious situations (Caruthers & Rodrigues 2009). On the bright side, although funding levels remain inadequate, REDDS (Reducing Emissions from Deforestation and Forest Degradation) Plus contains provisions whereby indigenous people who forego current consumption for the greater good, for example by not cutting down Amazonian rainforests, are compensated. This is an approach that could be used to preserve lands in their current form, provided that indigenous people are compensated for not using them now (TWN 2012).

Defining Green Social Work

Green social work as a form of practice rooted in enhancing the wellbeing of people and their environments becomes relevant to the realisation of environmental justice. Green social workers base their interventions on the idea that nation-states that are currently members of the United Nations (UN) are already committed to environmental justice because each member state is responsible for providing resources that mitigate risk for citizens endeavouring to procure the necessities of life because they have signed the Universal Declaration of Human Rights (UDHR). Under the UDHR's Articles 22 to 27, states are committed to enabling their citizens to ensure their wellbeing as far as it is reliant upon food, clothing, shelter, education and health resources. Additionally, various other treaties and protocols have promoted the idea of environmental justice as the right to live in a healthy environment, for example the Stockholm Declaration and the Rio Declaration. The right to a healthy and sustainable environment as an integral part of environmental justice has been endorsed by environmental movements and is integral to green social work. However, some states claim that they are unable to comply with the demands for food, clothing, shelter, education and health services as envisaged through the UDHR, even though on their own, these requirements stop short of the demand for environmental justice. State reluctance to guarantee the provision of even the basic human rights enshrined in the UDHR has been legitimated through contemporary fiscal crises. Under the guise of financial imperatives, politicians have condoned state abrogation of these responsibilities and their being shunted onto the shoulders of individuals and families. Moreover, this negation of people's entitlements has not been limited to countries in the Global South. It is also evident in Western states in the form of the public expenditure cuts that have been incorporated into the policies that have ushered in the 'age of austerity'. These cuts have decimated publicly funded welfare provisions and taken their toll on people's wellbeing. The plight of people in Greece, Spain and Portugal has covered acres of newsprint and megabytes of prime-time television coverage. . In contrast, the bankers whose inadequate governance of the financial institutions initially created the financial crisis, are rarely penalised and carry on maintaining their high standards of living (Jordan and Drakeford 2012).

Green social workers aim to promote environmental justice and resist environmental injustice by helping people mobilise and organise activities that protect their physical environment. Consequently, green social work practice affirms human rights and social justice to enhance the wellbeing of people and the environment for today and the future. Its concern to deal with forms of environmental degradation that undermine people's wellbeing provides the rationale that justifies social workers' involvement in environmental issues (Dominelli, 2011, 2012a). Green social work adopts a political stance in that it recognises that power relations shape human interactions and that these are rooted in an ethics of care whereby people care for one another and the environment in sustainable ways to ensure that all living things will survive now and for generations to come. Green social workers are critical of the unsustainable forms of development encapsulated in the contemporary neoliberal capitalist mode of production, reproduction and consumption because these exploit the environment and people's labour and perpetuate social and environmental injustice. Dominelli (2012a: 25) defined green social work as:

> a form of holistic professional social work practice that focuses on the: interdependencies amongst people, the social organisation of relationships between people and the flora and fauna in their physical habitats; and the interactions between socio-economic and physical environmental crises and interpersonal behaviours that undermine the well-being of human beings and planet earth. It proposes to address these issues by arguing for a profound transformation in how people conceptualise the social basis of their society, their relationships with each other, living things and the inanimate world. It does so by: tackling structural inequalities including the unequal distribution of power and resources; eliminating poverty and various 'isms'; promoting global interdependencies, solidarity and egalitarian social relations; utilising limited natural resources such as land, air, water, energy sources and minerals for the benefit of all rather than the privileged few; and protecting the earth's flora and fauna. The aim of green social work is to work for the reform of the socio-political and economic forces that have a deleterious impact upon the quality of life of poor and marginalised populations, secure the policy changes and social transformations necessary for enhancing the well-being of people and the planet today and in the future and advance the duty to care for others and the right to be cared by others.

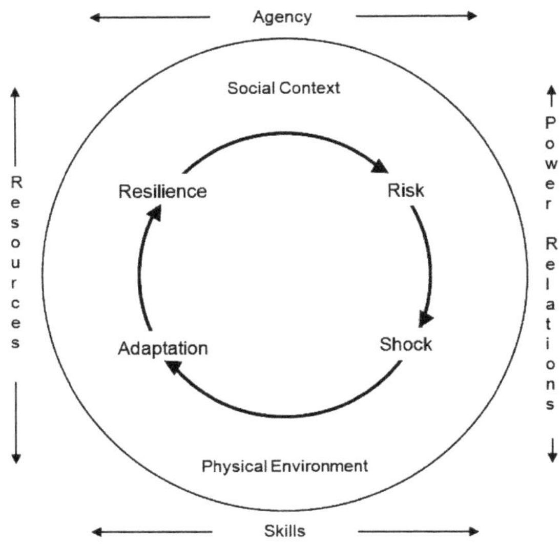

Figure 10.3 Green social work interventions

Its interventions are holistic, tackle structural forms of oppression, environmental degradation and injustice to empower people.

Green social workers are concerned about:

- Human rights violations
- Environmental degradation
- An economic system that perpetuates inequalities and lacks corporate accountability
- The neglect of cultural diversity and the non-affirmation of aboriginal and indigenous lifestyles
- Lack of people-friendly localities and supportive community relationships, especially those associated with unsustainable economic developments
- Lack of provision for the health and social services that promote the wellbeing of people
- Lack of care for the physical environment
- Lack of recognition of the interdependency among peoples and between people and the geo/ecosystem; and
- Lack of resilient-built infrastructures, resources and communities (Dominelli 2012a)

These concerns are linked to struggles for environmental justice while practitioners undertake the following tasks (Dominelli 2012a):

- Supporting people in affirming their human, social and environmental rights

- Supporting people in protecting the environment; enhancing the wellbeing of humans and the ecosphere; and obtaining environmental justice
- Mobilising people in various partnerships and alliances to promote residents' and the earth's wellbeing; and
- Empowering marginalised individuals, communities and groups, especially in influencing commercial and other institutional routines and policy/ decision-makers

In conducting their work, green social workers uphold the duty to care for and be cared by others, including the duty to care for Planet Earth. Green social workers endorse a holistic approach that envisages a unity between people and their physical, social, political, economic and cultural environments. These are part of one whole, and not separate from it.

Environmental degradation occurs in the course of meeting people's needs. Much of it could be reduced by more careful planning that draws upon holistic environmental assessments that enhance existing forms of resilience. Involving local people in decision-making processes and disaster reduction initiatives that promote sustainable development can avoid poor decisions including those that undermine environmental integrity. Additionally, such practices underpin the co-production of knowledge and partnerships that cross disciplinary divides and facilitate contact between local people and experts.

Top-down decision-making that ignores local knowledge can destroy environmental integrity, force the migration of large numbers of people and produce unforeseen social and environmental problems, for example the dead water and drought that has occurred following the construction of the Three Gorges Dam in China (Watts 2011). Such projects also highlight tensions between governments engaging in grandiose status-raising projects and the smaller, life-sustaining ones preferred by local people who wish to remain in their communities of origins rather than migrate elsewhere.

Social workers are active if invisible players in tackling social and environmental degradation associated with disasters and reconstructing damaged communities. A practitioner working in such situations assumes various roles that require understanding complex power dynamics and relationships in pursuing environmental justice. These include that of:

- facilitator
- coordinator
- community mobiliser (of people and systems)
- mobiliser of resources
- negotiator or broker between communities and different levels of government
- mediator between conflicting interests and groups, including gendered relations
- consultant to government and other agencies

- advocate for people's rights and entitlements
- educator, giving out information about how to access relief aid and avoid diseases that can erupt following a disaster
- trainer, particularly in how to respond effectively in mobilising local resources when disaster strikes
- cultural interpreter
- interdisciplinary translator
- therapist helping people deal with the emotional consequences of disaster
- protector of the ecosystem/physical environment (Dominelli, 2009, 2012a)

Intervening in disaster situations embedded in environmental degradation can be complicated and messy – physically, socially and emotionally. Clarity about cultural expectations and norms that can militate against fairness and clearing rubble when the physical environment and infrastructures are damaged become relationship-building skills. In helping people piece their lives together, green social workers in such disasters undertake the following tasks:

- assessing need
- co-ordinating and delivering goods and services
- assisting in family reunification
- supporting individuals and communities to rebuild their lives, develop resilience and build capacity to minimise future risks, and
- advocating, lobbying and mobilising for changes that safeguard the environment and prevent future disasters

For green practitioners, supporting physically and psychosocially injured people and repairing damaged environments become starting points for longer-term transformative initiatives that develop environmentally just, sustainable, life-enhancing forms of being and doing.

Conclusions

Green social work is essential for reducing the footprint that people leave on the environment, ensuring that all the earth's inhabitants have an equitable share of resources, protecting the earth's flora and fauna, and promoting sustainable forms of development that will not cost the earth or forfeit the future. Green social work is based on the idea of a citizenry that transcends particular locales to acknowledge the interdependencies between peoples and the realisation of peoples' human rights, social and environmental justice and citizenship entitlements. It encompasses a critique of consumerism, hyper-urbanisation, neoliberalism and expert-led solutions to the social challenges that currently face humanity. It contains a commitment to facilitating local residents' action (agency), valuing their knowledge and co-producing solutions alongside all other stakeholders with an interest in protecting people and the living environment from harm. In promoting and enhancing environmental justice, green social work practice is also good social work practice that draws on empowering values and skills. Hence, environmentally just, green social work practice lies at the heart of the profession.

Acknowledgements

This chapter draws on a lecture I gave when I was awarded the 2012 Katherine A Kendall Award in Stockholm, Sweden. I acknowledge the contributions to the insights developed here through research projects funded by two research councils in the UK, the ESRC (Internationalising Institutional Practices 2009 to 2012); and EPSRC (Built Infrastructures, Older People and Climate Change (BIOPICCC) 2009 to 2012).

References

Bullard, D. (1990) Dumping in Dixie. Race, Class and Environmental Quality. New York: Westview Press.

Brundtland, G. (1987) The World Commission on the Environment and Development. New York: United Nations.

Carruthers, D. and Rodriguez, P. (2009) Mapuche Protest, Environmental Conflict and Social Movement Linkage in Chile. Third World Quarterly, 30(4), 743–760.

Davis, M. (1998) Ecology of fear: Los Angeles and the Imagination of Disaster. New York: Metropolitan Books.

Dessler, A. and Parsons, E. (2006) The science and politics of climate change: A guide to the debate. Cambridge: Cambridge University Press.

Dominelli, L (2011) 'Climate Change: A Social Work Perspective', International Journal of Social Welfare, October, 2011, 20(4): 430–438.

Dominelli, L. (2012a) Green Social Work. Cambridge: Polity Press.

Dominelli, L. (2012b) Mind the Gap: Built Infrastructures, Sustainable Caring Relations and Resilient Communities in Extreme Weather Events, Australian Social Work, pp. 1–14, DOI:10.1080/0312407X.2012.708764.

Dominelli, L. (2012c) Gendering Climate Change: Debates, Policies and Action, in Research, Action and Policy: Addressing the Gendered Impacts of Climate Change (pp. 77–94), edited by M. Alston and K. Whittenbury. Sydney Australia: Springer.

Dupuy, J.P. (2005). Petite metaphysique des tsunamis [A Modest Philosophising about Tsunamis]. Paris: Seuil (Enlightened catastrophism).

Evans, W. and Manion, M. (Eds.) (2002) Minding the Machines: Preventing Technological Disasters. New York: Prentice Hall.

Giddens, A. (2009) The Politics of Climate Change. Cambridge: Polity.

IFRC (International Federation of the Red Cross and Red Crescent Societies) (2009) World Disasters Report: Focus on Displacement and Migration. Geneva: IFRC.

Jordan, B. and Drakeford, M. (2012) Social Work and Social Policy under Austerity. Palgrave: Macmillan.

Kroll, L. & Fass, A. (2011). The World's Billionaires. Forbes Magazine, Special Report, 8 March.

Klinenberg, E. (2002) Heat Wave: A Social Autopsy of Disaster in Chicago. Chicago: University of Chicago Press.

Laugier, S. and Paperman, P. (Eds.) (2006) Le souci des autres: Ethique et politique du care [The Care of Others: Ethics and Policy]. Paris: Editions de l' EHESS.

Lavalette, M. and Jones, C. (2011) Voices From the West Bank: Young people living under occupation. Liverpool: Bookmarks.

Lovins, A. (2011) Soft Energy Paths for the 21st Century. Rocky Mountain Institute Document E11–09. Retrieved from http://www.rmi.org/Knowledge-Center/Library/2011–09_GaikoSoftEnergyPaths

Newsweek (2012). 4–11 June 2012, 49.

Oven, K.J., Curtis, S.E., Reaney, S., Riva, M., Ohlemüller, R., Syewart, M.G., Dunn C.E., Nodwell, S., Dominelli, L., and Holden, R. (2011) Climate change and health and social care: Defining future hazard, vulnerability and risk for infrastructure systems supporting older people's health care in England. Journal of Applied Geography, 33(1), 16–24, doi:10.1016/j.apgeog.2011.05.012.

Pearce, F. (2012) Could a Climate Change Set Off Volcanoes and Quakes? Environment 360: Opinion, Analysis, Reporting and Debates. 5 May. Retrieved from http://e360.yale.edu/feature/could_a_changing_climate_set_off_volcanoes_and_quakes/2525/.

Pyles, L. (2007) Community Organising for Post-Disaster Development: Locating Social Work. International Social Work, Special Issue on Disasters, 53(5): 321–333.

Robine, J.M., Cheung, S., Le Roy, S., Van Oyen, H., Griffiths, C., Michel, J.P., and Herrmann, R. (2007) Death Toll Exceeded 46,000 in Europe during the Summer of 2003. Comptes Rendues Biologies, 331(2), 171–178.

Schmidtlein, M., Deutsch, R., Piegorsch, W., and Cutter, S. (2008) A sensitivity analysis of the social vulnerability index. Risk Analysis, 28(4), 1099–1114.

Sobie, J. (2011) TEPCO Tumbles on Fresh Bankruptcy Fears. Financial Times, 6 June. Retrieved from http://www.ft.com/cms/s/0/4639a818–9020–11e0–85a0–00144feab49a. html#axzz2FdzpLoOw.

Toole, M. (2006) Forced Migrants, Refugees and Internally Displaced Persons, in Social Injustice and Public Health (pp. 190–204), edited by B. Levy and V. Sidel. Oxford: Oxford University Press.

Third World Network, TWN (2012). Doha News Update, No. 27, 19 December.

War Child (2012) Congo's Conflict Minerals. Retrieved from http://www.warchild.org.uk/issues/congo-conflict-minerals?gclid=CLS44NegqrQCFal7Q godhDUAPQ.

Watts, J. (2011) China Warns of 'Urgent Problems' Facing Three Gorges Dam. The Guardian, 20 May. Retrieved from http://www.guardian.co.uk/world/2011/may/20/three-gorges-dam-china-warning.

Wisner, B., Blaikie, P., Cannon, T., and Davis, I. (2004) At Risk: Natural Hazards, People's Vulnerability, and Disaster. London: Routledge.

Zeller, T. (2012) Flood Threat to Nuclear Plants Covered Up by Regulators, NRC Whistleblower Claims. Huffington Post, 9 May. Retrieved from http://www.huffingtonpost.com/2012/09/14/flood-threat-nuclear-plants-nrc_n_1885598.html.

Appendix[1] The Global Agenda for Social Work and Social Development Commitment to Action[2]

Our Commitments

As social workers, educators and social development practitioners, we witness the daily realities of personal, social and community challenges. We believe that now is our time to work together, at all levels, for change, for social justice, and for the universal implementation of human rights, building on the wealth of social initiatives and social movements.

We, the International Federation of Social Workers (IFSW), the International Association of Schools of Social Work (IASSW), and the International Council on Social Welfare (ICSW), recognise that the past and present political, economic, cultural and social orders, shaped in specific contexts, have unequal consequences for global, national and local communities and have negative impacts on people. Specifically, we recognise that:

- the full range of human rights are available to only a minority of the world's population
- unjust and poorly regulated economic systems, driven by unaccountable market forces, together with non-compliance with international standards for labour conditions and a lack of corporate social responsibility, have damaged the health and wellbeing of peoples and communities, causing poverty and growing inequality[3]
- cultural diversity and the right to self-expression facilitate a more satisfactory intellectual, emotional, moral and spiritual existence[4], but these rights are in danger due to aspects of globalisation which standardise

1 Reproduced with acknowledgment of ICSW, IASSW and IFSW.

2 This version is provided by the former President of IFSW David N Jones. He is also one of the social work representatives, responsible for this document.

3 International Labour Organisation (ILO) (2008) Declaration on Social Justice for a Fair Globalization.

4 United Cities and Local Governments (2004) – Agenda 21 for culture.

and marginalise peoples, with especially damaging consequences for indigenous and first nation peoples[5]

- people live in communities and thrive in the context of supportive relationships, which are being eroded by dominant economic, political and social forces
- people's health and wellbeing suffer as a result of inequalities and unsustainable environments related to climate change, pollutants, war, natural disasters and violence to which there are inadequate international responses

Consequently, we feel compelled to advocate for a new world order which makes a reality of respect for human rights and dignity and a different structure of human relationships[6].

Therefore:

We commit ourselves to supporting, influencing and enabling structures and systems that positively address the root causes of oppression and inequality.

We commit ourselves wholeheartedly and urgently to work together, with people who use services and with others who share our objectives and aspirations, to create a more socially-just and fair world that we will be proud to leave to future generations.

We will prioritise our endeavours to these ends.

We intend during the period 2012–2016 to focus our efforts on the following areas:

- Promoting social and economic equalities
- Promoting the dignity and worth of peoples
- Working toward environmental sustainability
- Strengthening recognition of the importance of human relationships

The commitments are guided by and consistent with our core statements on the definition of social work[7] and the ethical principles of social work[8].

Our Organisations

This Global Agenda is the product of a three year collaborative initiative undertaken by three international organisations representing social work practice, social work education and social development. All three of these international bodies were founded in 1928 and have held formal consultative status for many

5 UNESCO Universal Declaration on Cultural Diversity (2001).

6 Global Agenda for Social Work and Social Development.

7 International Definition of Social Work (under review).

8 Ethics in Social Work, Statement of Principles.

decades with the United Nations Economic and Social Council and other UN and related agencies.

The International Federation of Social Workers is the global federation of national social work organisations in 90 countries representing over 750,000 social workers[9].

The International Association of Schools of Social Work is an international community of schools and educators in social work, promoting quality education, training and research in the theory and practice of social work, administration of social services and formulation of social policies. IASSW speaks on behalf of 2,000 schools of social work and 500,000 students[10].

The International Council on Social Welfare is a global, non-governmental organisation which represents tens of thousands of organisations around the world that are actively involved in programmes to promote social welfare, social development and social justice[11].

In preparing the Global Agenda, we have consulted extensively with social workers, social work educators and social development practitioners. In 2010, at our joint conference in Hong Kong, the Global Agenda received overwhelming support from approximately 3,000 delegates. Consultation has culminated in the following specific Commitments to Action.

Our Role In Promoting Social And Economic Equalities

Target group: The United Nations and other international agencies

We commit to support, influence and promote global initiatives aimed at achieving social and economic equality. We will accomplish this by using and strengthening our established relationships with the UN system and other international agencies. We will support the achievement of the Millennium Development Goals. Our major focus is to prepare for the post-2015 development agenda[12], which includes, for example, the social protection floor initiative[13]; decent work and international

9 www.ifsw.org.

10 www.iassw-aiets.org.

11 www.icsw.org.

12 UN General Assembly (2011) Accelerating progress towards the Millennium Development Goals – Annual report of the Secretary-General.

13 The term "social protection" is used to mean protection provided by social security systems in the case of social risks and needs. Social protection is often interpreted as having a broader character than social security (including, in particular, protection provided between members of the family or members of a local community). http://www.social-protection.org.

labour standards[14]; the WHO initiative on the social determinants of health[15]; and education for all[16].

We will strive with others for a people-focused global economy that is regulated to protect and promote social justice, human rights and sustainable development.

Target Group: Communities and other partners

We will support and work in collaboration with others for the development of strong local communities that promote the sustainable social wellbeing of all their members. Our major focus is to strengthen the capacity of communities to interact with their governments to extend social and economic development.

Target Group: Our own organisations

We will work within our own organisations to promote education and practice standards in social work and social development that enable workers to facilitate sustainable social development outcomes.

Our Role in Ensuring the Dignity and Worth of the Person

Target group: The United Nations and other international agencies

We seek universal implementation of the international conventions and other instruments on social, economic, cultural and political rights for all peoples,

14 ILO (2008) Decent work agenda 'Decent work sums up the aspirations of people in their working lives. It involves opportunities for work that is productive and delivers a fair income, security in the workplace and social protection for families, better prospects for personal development and social integration, freedom for people to express their concerns, organise and participate in the decisions that affect their lives and equality of opportunity and treatment for all women and men.' (See ILO (2008) Declaration on Social Justice for a Fair Globalization above.)

15 World Health Organisation (WHO) (2008) Commission on Social Determinants of Health 'The social determinants of health are the conditions in which people are born, grow, live, work and age, including the health system. These circumstances are shaped by the distribution of money, power and resources at global, national and local levels, which are themselves influenced by policy choices. The social determinants of health are mostly responsible for health inequities – the unfair and avoidable differences in health status seen within and between countries.'

16 UNESCO (2000) Education for All: Meeting our Collective Commitments.

including, among others, the rights of children[17]; older people[18]; women[19]; persons with disabilities[20] and indigenous peoples[21], and the end to discrimination on the grounds of race and sexual orientation[22].

We will promote social strategies that build cohesive societies and remove the seeds of conflicts. We seek renewed commitment to the peaceful prevention and resolution of conflict and adherence to international agreements which can reduce violence and its consequences[23].

We will work with our partners to challenge violent state responses to actions by people in defence of their rights.

We will advocate for the right of people to move between and within countries and for the right of documented and undocumented migrants to have access to social services.

We will support measures to reduce and eliminate human trafficking[24].

Target Group: Communities and other partners

We will support communities and organisations that are engaged in the prevention of violent conflict both between and within countries.

We will support actions taken by partner organisations to enhance the rights of internal or trans-national migrants.

We will partner with international, regional and national entities to combat human trafficking.

We will partner with international, regional and national entities to promote respect for cultural identity.

---

17 United Nations (1989) 'Convention on the Rights of the Child'.

18 Global Action on Aging (2009) 'Toward a Human Rights Convention for Older Persons'.

19 United Nations (1979) 'Convention on the Elimination of All Forms of Discrimination against Women'.

20 United Nations (2006) 'Convention on the Rights of Persons with Disabilities'.

21 United Nations (2007) 'Declaration on the Rights of Indigenous People'.

22 UN Human Rights Council (2011) 'Human rights, sexual orientation and gender identity'.

23 For example United Nations Office for Disarmament Affairs (2011) 'Towards an Arms Trade Treaty' and United Nations (1997) 'Convention on the Prohibition of the Use, Stockpiling, Production and Transfer of Anti-Personnel Mines and on their Destruction'.

24 United Nations (2003) 'Protocol to Prevent, Suppress and Punish Trafficking in Persons, Especially Women and Children', supplementing the United Nations Convention against Transnational Organized Crime – see also http://www.unodc.org/unodc/en/human-trafficking/what-is-human-trafficking.html.

Target Group: Our own organisations

We will ask members of our three organisations to work together to ensure that legislation and practices do not prevent or limit the right to undertake legitimate, democratic social development activities.

We will promote respect for diversity and advocate for education and training programmes that prepare social work and social development practitioners for ethical and informed interventions[25], for example in relation to gender, sexual orientation and culture.

We will work to promote education and practice standards that prepare social work and social development practitioners to be effective in the areas of conflict management, human trafficking and responding to the consequences of migration.

Our Role in Promoting Sustainable Communities and Environmentally Sensitive Development

Target group: The United Nations and other international agencies

We commit to aligning our activities and programmes to development initiatives that are sustainable and which integrate the human dimension. We will strengthen our established relationships with the UN and other international agencies in order to reaffirm our support for initiatives aimed at protecting the natural environment. Our major focus will be on the Rio +20 process[26], World Urban Forum[27] and the post-2015 development agenda, including attention to disaster prevention and management[28].

Target Group: Communities and other partners

We will promote community capacity building in responding to environmental challenges and human and natural disasters (e.g. flooding, water degradation, food insecurity).

25 See for example IASSW/IFSW (2004) 'Global Standards For The Education And Training Of The Social Work Profession'.

26 United Nations (2011) 'Conference on Sustainable Development (UNCSD) (Rio +20).

27 United Nations (2011) 'World Urban Forum 6'.

28 For example United Nations DESA (2011) 'Settlement Planning: UN-HABITAT in disaster and conflict contexts' and 'Office for the Coordination of Humanitarian Affairs' (OCHA).

Target Group: Our own organisations

We will promote within our own organisations standards in education and practice that facilitate sustainable social development outcomes, including the prevention, mitigation and response to disasters. We will encourage and facilitate research into the social work role in relation disasters and environmental challenges.

Our Role in Promoting Wellbeing through Sustainable Human Relationships

Target group: The United Nations and other international bodies

We commit to working with the UN and other international bodies to promote the importance of sustainable, interdependent communities to achieve social development and wellbeing. We will advocate for methodologies which support community empowerment.

Target Group: Communities

We will work in collaboration with others to promote strong inclusive communities that enable all members to participate and belong. We will promote policies aimed at 'social integration and cohesion' as a means for achieving the economic and social wellbeing of all persons, including older people and persons with disabilities, mental health needs and/or learning difficulties.

Target Group: Our own organisations

We will promote within our own organisations the importance of reducing social isolation and building social interaction through social relationships, embedding the principles of social cohesion, development and inclusion in any published policies and standards relating to education and practice.

Ensuring an Appropriate Environment for Practice and Education

We will promote the development, dissemination and exchange of knowledge between all social professionals through established and innovative channels of communication.

We will support the development of regional research centres for social work/development and encourage research to identify work environments which promote positive outcomes in social work and social development.

Conclusion

The three organisations are firmly committed to implementation of The Agenda Commitments in a planned programme from 2012–2016, aiming to work alongside others who share these commitments. We will establish an implementation plan with arrangements for monitoring and review. We will provide regular reports on implementation to the global community of social workers, social work educators and social development workers – and beyond.

6 January 2012

Index

For Product Safety Concerns and Information please contact our EU representative GPSR@taylorandfrancis.com
Taylor & Francis Verlag GmbH, Kaufingerstraße 24, 80331 München, Germany

www.ingramcontent.com/pod-product-compliance
Ingram Content Group UK Ltd.
Pitfield, Milton Keynes, MK11 3LW, UK
UKHW011013180926
475701UK00004B/722